Editors/Authors:

Jiaxin Chen – Authors

Zhongying Jin – Editor

Feng Wang – Illustrator

Publisher:

Golden Abacus Pty Ltd

ABN: 776 214 795 34

2/18 Fraser Street, Glen Waverley VIC 3150

www.vsuccess.com.au

Trade Enquiries:

Phone: +61 422 281 978

Email: info@vsuccess.com.au

ISBN: 978-0-6450402-7-2

2nd Edition: December 2024

**Dear Students and Parents,**

Thank you for purchasing *Core Mathematics* series book. This book mainly follows the Australian Curriculum, integrated with other Maths Curriculum across the world. Each module will have basic questions and challenging questions, allowing students to make learning progress by solving different sets of problems step by step.

*Core Mathematics* is for those advanced students who are interested in learning mathematics and hope to discover their potentials. Many topics in this book needed to be taught and presented by teachers in a professional way.

We hope that through learning a series of *Core Mathematics* workbooks, every student can not only build their understanding of essential math knowledge, but also develop creative and critical thinking skills. Therefore, students can be more prepared for future learning and gain confidence when dealing with challenging problems and unfamiliar situations in real life.

I hope *Core Mathematics* will make you feel enjoyable and wish you all have a wonderful learning journey.

Jiaxin Chen

# 目录

## Module 1 认识数字 Introduction to numbers

### ❖ 认识 0 - 9 Knowing 0 to 9

Numbers play an important role in our lives because our daily life events involve various numbers, such as shopping, studying, entertainment or working.

Today we will learn the ten numbers from 0 to 9 as they can be used to form any numbers in the world, and they will be with us for a long time.

| | | |
|---|---|---|
| 0 |  | |
| 1 |  |  |
| 2 |  |  |
| 3 |  |  |
| 4 |  |  |
| 5 |  |  |
| 6 | 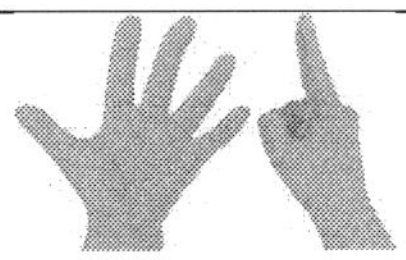 |  |
| 7 | 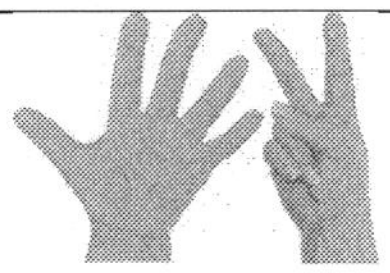 |  |
| 8 | 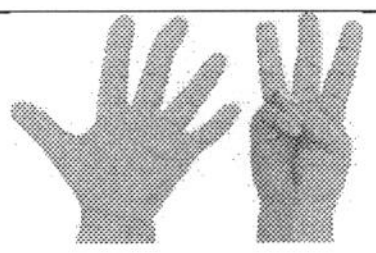 |  |
| 9 | 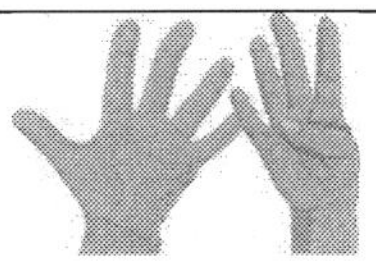 |  |

❖ 数一数 **Counting numbers**

From the table above, we can see that fingers can be used to represent different numbers and amounts. Please find which number is represented by each of the following amounts.

| | |
|---|---|
|  | |
|  | |
|  | |
|  | |
|  | |
|  | |

❖ 序数 **Ordinal numbers**

Ordinal numbers can give us the exact position of an object. In other words, they tell the position of an object in relation to other objects rather than the quantity.

Ordinal numbers are also commonly used in daily life. Sometimes, we might not notice them every time we use it.

For example, the graph below shows the five athletes competing in a marathon race. We can use ordinal numbers to describe their performance.

The numbers $1^{st}$, $2^{nd}$, $3^{rd}$, $4^{th}$, $5^{th}$ can tell the ranking of different athletes in the race.

**Module 1 Practice Lesson 1**

1. Find the number of items in each picture. Then match it with the correct quantity.

3

6

1

8

7

9

2

4

5

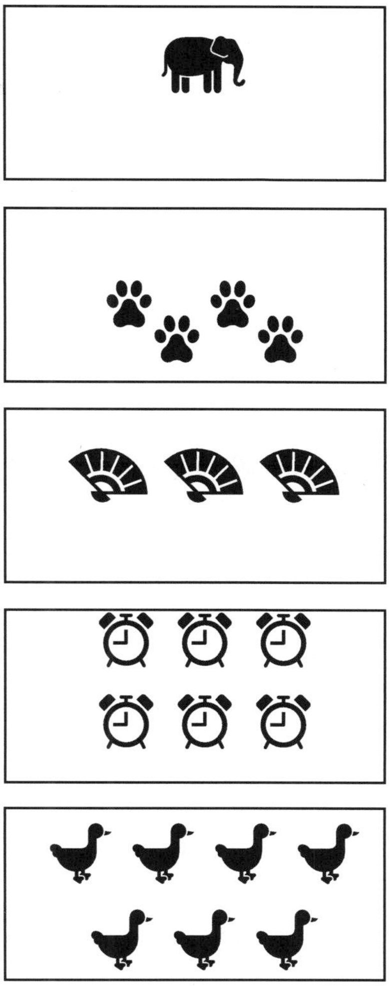

2. Tick the correct amount by looking at the fingers shown in each row of the table below.

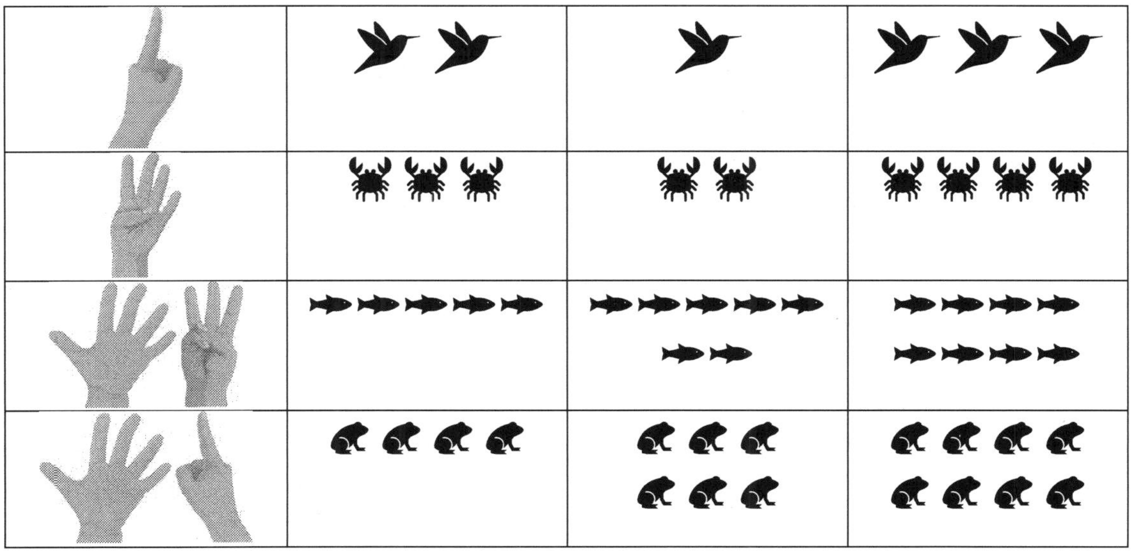

3. Circle the correct number of objects by looking at the numbers given.

| 3 | 4 |
|---|---|
|  |  |

| 6 | 7 |
|---|---|
|  |  |

4. Complete the table below by counting the number of each object.

|  |  |  |  |  |  |  |  |  |  |
|---|---|---|---|---|---|---|---|---|---|
| | | | | | | | | | |

## Module 1 Practice Lesson 2

1. Match.

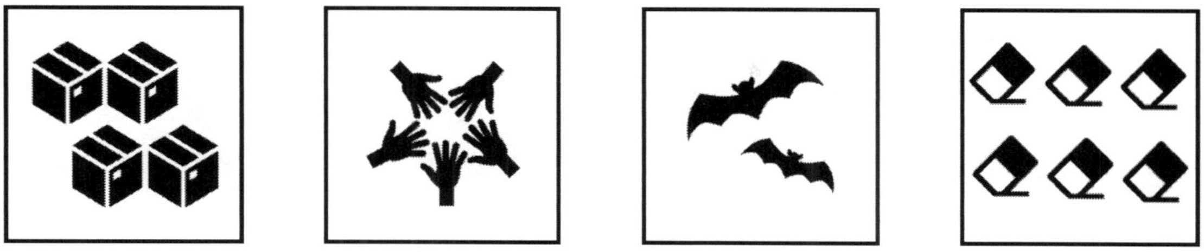

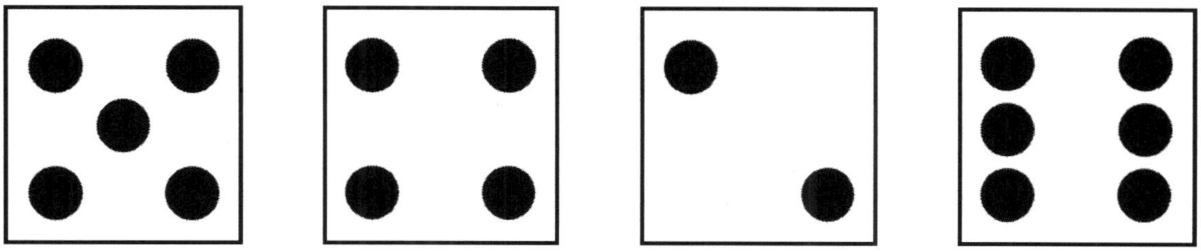

2. Count the number of each item first and then draw the same number of circles.

| | |
|---|---|
|  |  |
| | |

| | |
|---|---|
|  |  |
| | |

3. Look at each pattern below and draw the missing shapes in the boxes provided.

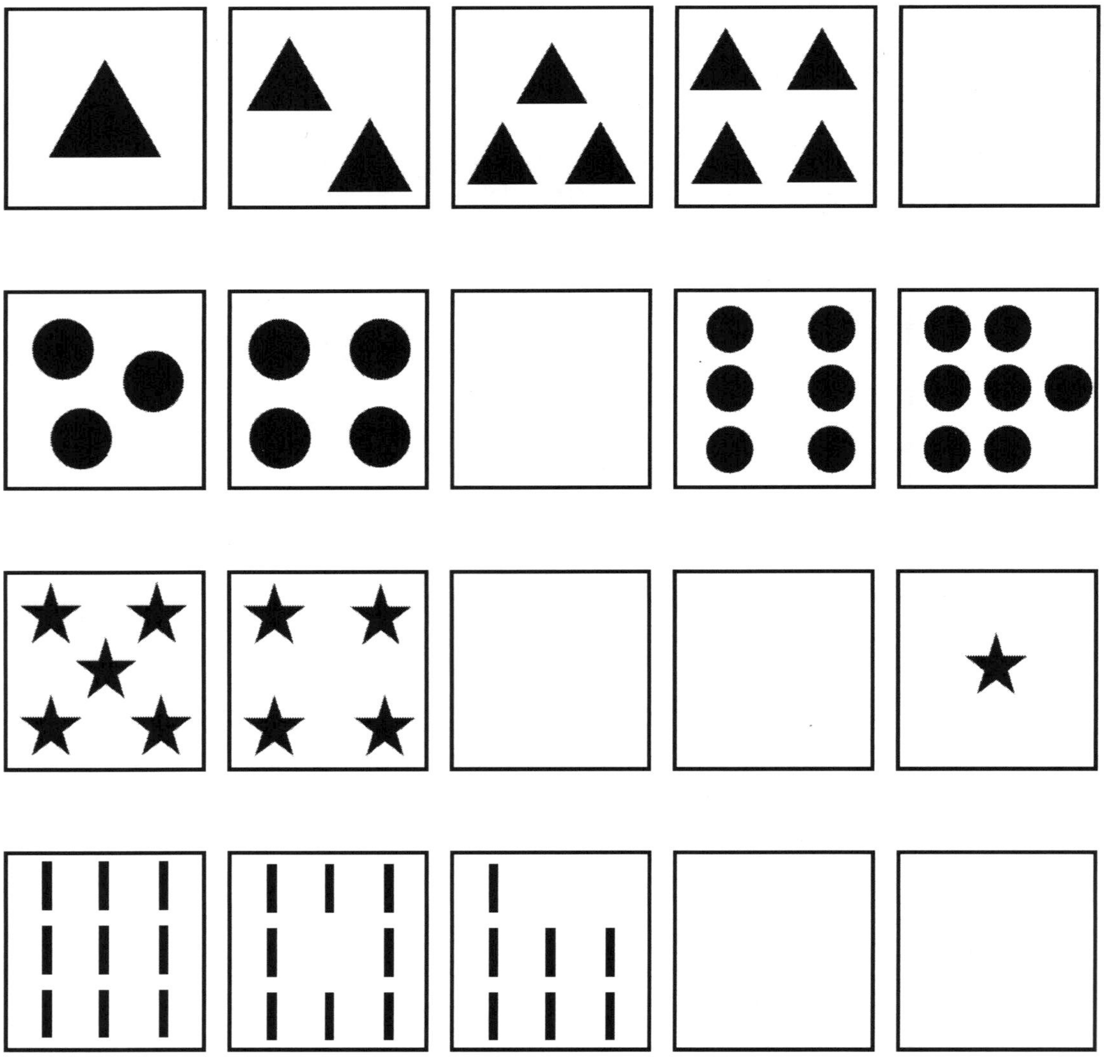

4. Write down the ranking using ordinal numbers.

## Module 1 Practice Lesson 3

1. Connect the objects with the same amount.

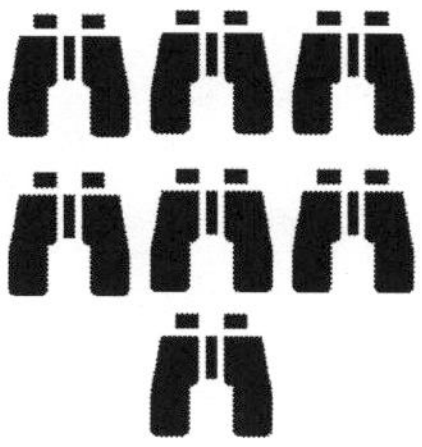

2. Circle the correct number to describe the amount in each picture.

| | |
|---|---|
|  |  |
| 4 5 6 7 8 9 | 4 5 6 7 8 9 |

| | |
|---|---|
|  |  |
| 4 5 6 7 8 9 | 4 5 6 7 8 9 |

3. For each of the following objects, find its amount and connect with the correct number.

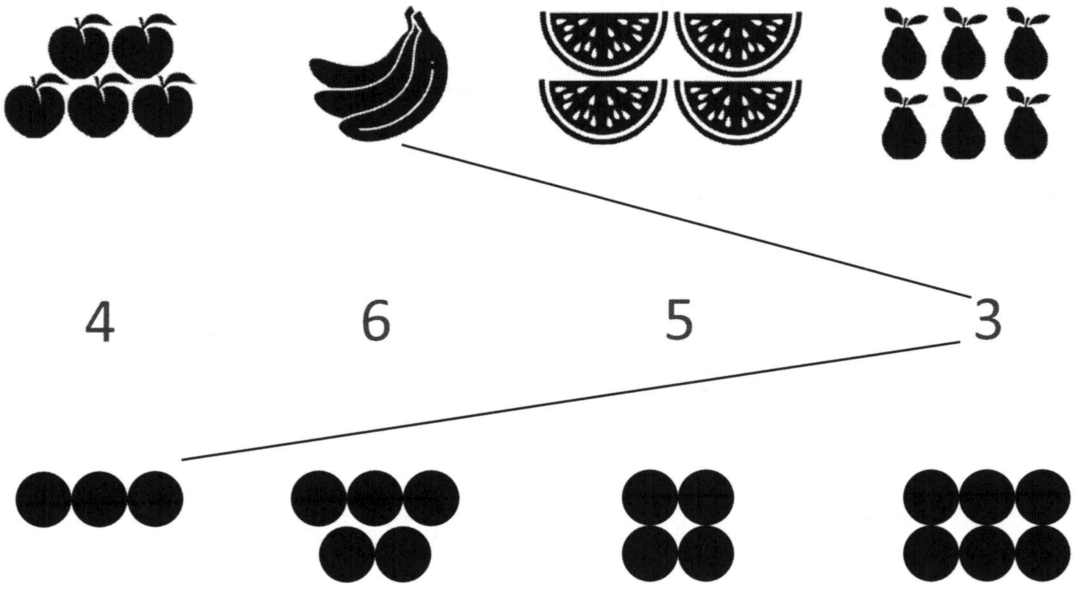


4. Trace the following numbers.

0 1 2 3 4 5 6 7 8 9

0 1 2 3 4 5 6 7 8 9

5. The following diagram shows the running results of a school carnival. Can you figure out the first, second and third places?

Joe is in the ________________ place.

Ben is in the ________________ place.

Tim is in the ________________ place.

## Module 1 Practice Lesson 4

1. Connect the numbers in order using straight lines.

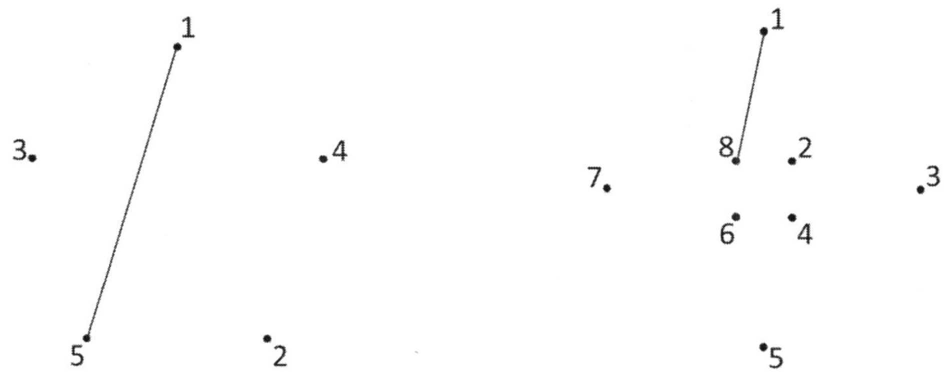


2. There are some people waiting in a queue at the counter of a supermarket.

a) How many people are in the queue?

b) If Anna is the first in the queue, where is Iris?

c) If Anna is the last in the queue, where is Ben?

d) If Kat is the last in the queue, who is the third last in the queue?

3. Colour the correct number of shapes with your favourite colour.

| | | |
|---|---|---|
| 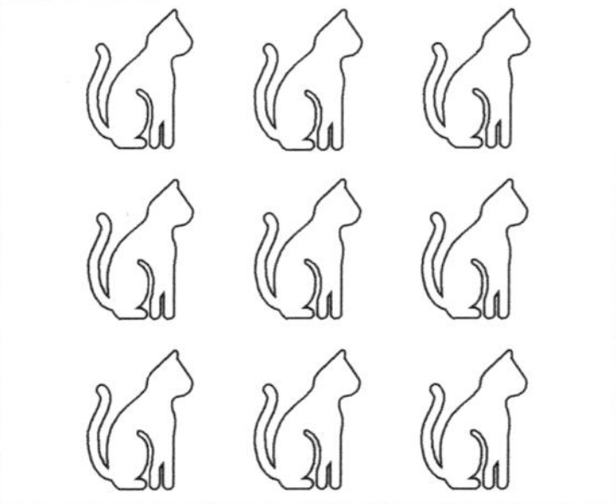 |  | 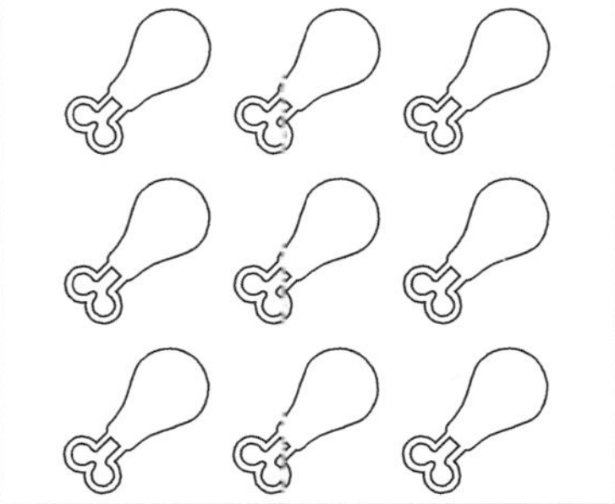 |
| 5 | 6 | 8 |

| | | |
|---|---|---|
| 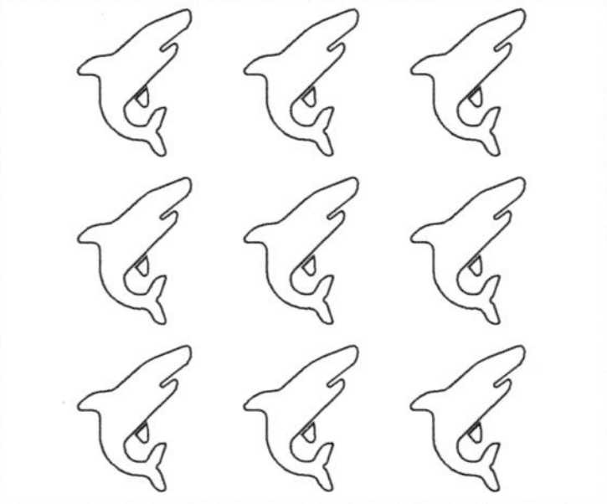 |  | 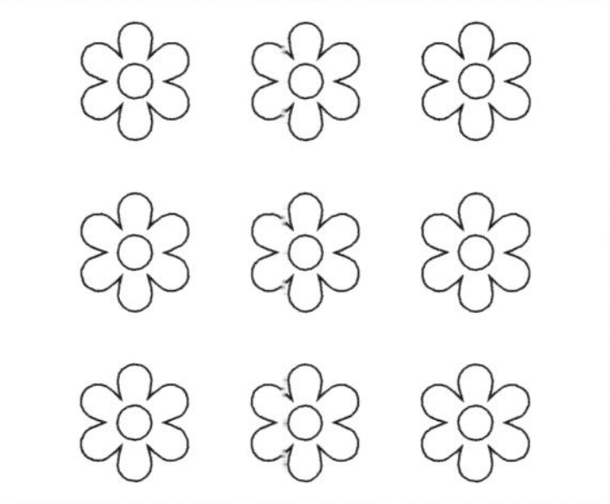 |
| 9 | 4 | 7 |

4. Look at each pattern below and fill the missing numbers.

| | | | | | |
|---|---|---|---|---|---|
| | | 2 | | | 5 |

| | | | | | |
|---|---|---|---|---|---|
| 7 | | | | 3 | |

| | | | | | |
|---|---|---|---|---|---|
| | | 7 | | 5 | |

| | | | | | |
|---|---|---|---|---|---|
| | 3 | | | 6 | |

## Module 2 比较 Compare

### ❖ 比较多少 Compare quantity

When we compare two quantities, we need to find which one has more amount and which one has less amount.

In maths, we can use symbol to compare, such as greater than ( > ) and less than ( < ).

The best way to remember 'the greater than/less than symbols' is to imagine them as little crocodiles. The numbers on each side of the symbol represent different amounts of fish. The crocodile always wants to eat the larger number of fish. So its mouth always open towards the larger number.

5 > 1

2 < 4

### ❖ 比较长短 Compare length

Same as comparing quantities, the crocodile will face to the longer length.

B > A

A < B

❖ 比较轻重 **Compare weight**

Same as comparing quantities, the crocodile will face to the heavier object.

Whale > Crab

Cat < Cow

❖ 比较空间大小 **Compare size**

Same as comparing quantities, the crocodile will face to the shape with a larger size.

A > B

A < B

## Module 2 Practice Lesson 1

5. Connect the pictures with the same quantities.

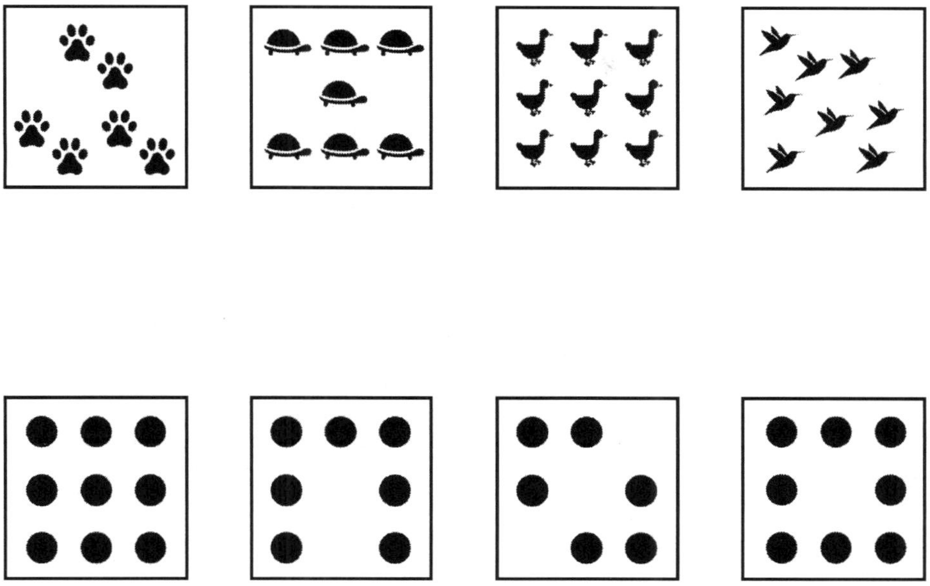

6. Put a tick ( √ ) next to the object with the most quantity.

7. Put a tick ( √ ) next to the object with the least quantity.

8. In each picture below, put a tick ( √ ) next to the object with the most quantity.

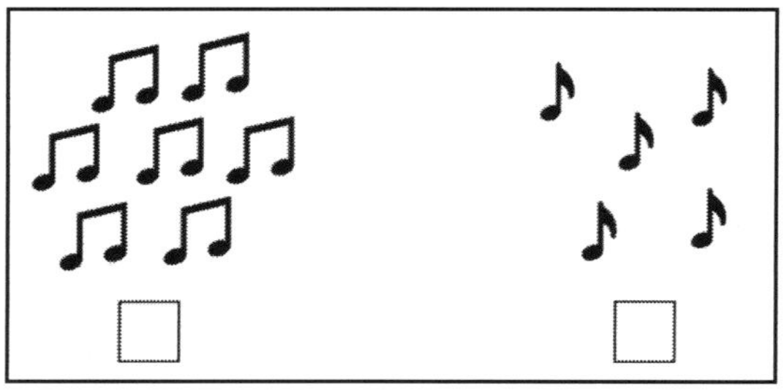

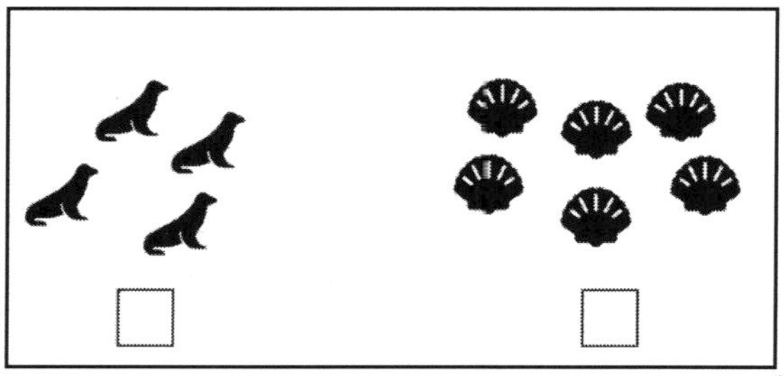

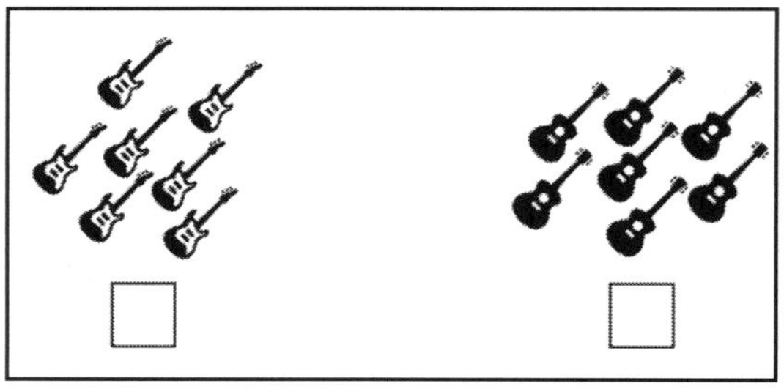

9. In each picture below, put a tick ( √ ) next to the object with the least quantity.

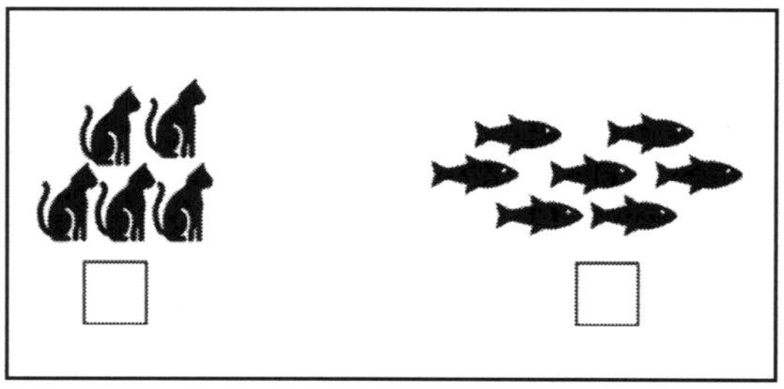

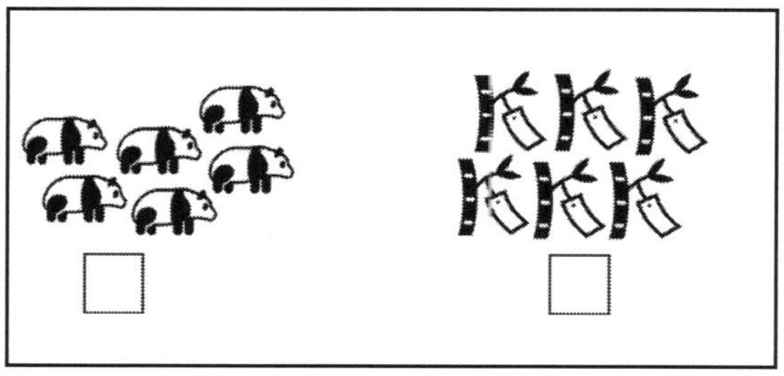

10. Draw as many circles ○ as triangles △.

## Module 2 Practice Lesson 2

5. Put a tick ( √ ) next to the longer length.

6. Put a tick ( √ ) next to the shorter length.

7. In each set, compare and decide which object is heavier. Put a tick ( √ ) to show your answer.

8. In each set, compare and decide which shape has a smaller size.
Put a tick ( √ ) to show your answer.

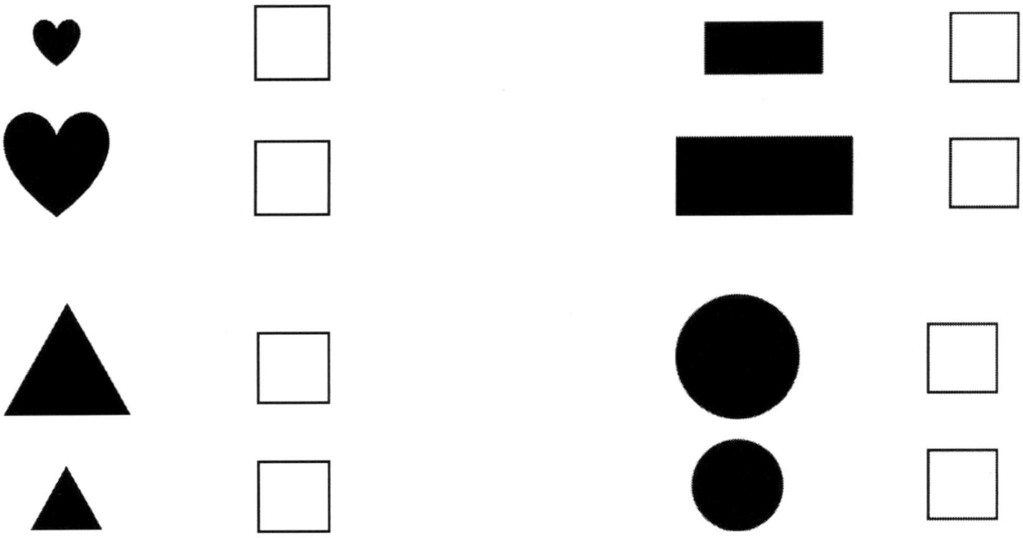

9. Put a tick ( √ ) next to the container that has more water.

10. Put a tick ( √ ) next to the container with less amount of water.

11. The number of circles ○ is 3 more than the number of triangles △. How many circles should be drawn?

## Module 2 Practice Lesson 3

6. Compare the following objects and write ">", "<" or "=".

| | | |
|---|---|---|
| 3 ○ 3 | 2 ○ 3 | 4 ○ 1 |

7. Trace the following symbols.

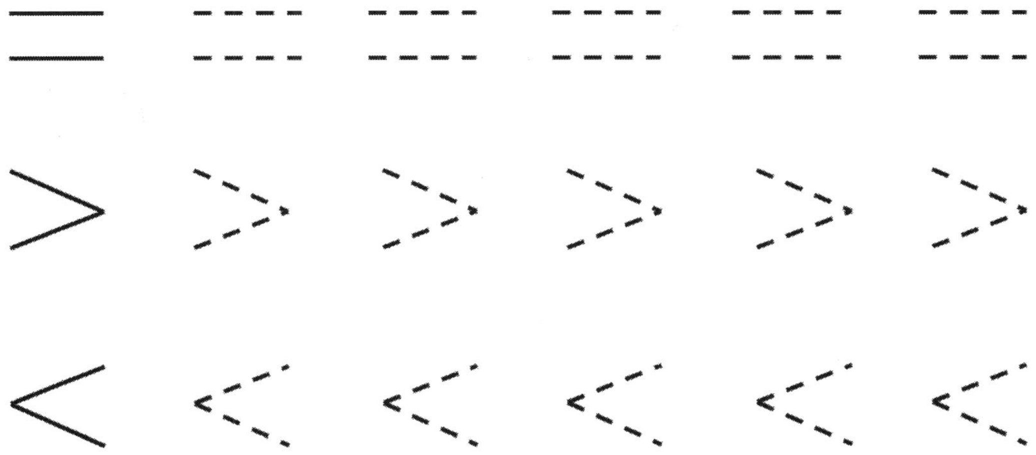

8. Compare the weights of each set of objects and write ">", "<" or "=".

9. Compare each set of numbers and write “>”, “<” or “=”.

| | |
|---|---|
| 3 □ 4 | 5 □ 8 |
| 5 □ 2 | 6 □ 7 |
| 7 □ 6 | 6 □ 5 |
| 9 □ 7 | 1 □ 3 |

10. What numbers can you put to make each statement correct? Answers may vary.

| | |
|---|---|
| 3 < □ | 4 < □ |
| 1 > □ | 7 > □ |
| 2 < □ | 6 < □ |
| 3 > □ | 9 > □ |

## Module 2 Practice Lesson 4

5. Fill the missing boxes to make each statement true. The first one has been done for you.

| 3 | | 5 | 7 | | 6 |
|---|---|---|---|---|---|
| 3 | $<$ | 5 | □ | $<$ | □ |
| 5 | $>$ | 3 | □ | $>$ | □ |

| 8 | | 7 | 4 | | 9 |
|---|---|---|---|---|---|
| □ | $<$ | □ | □ | $<$ | □ |
| □ | $>$ | □ | □ | $>$ | □ |

6. Order the following numbers from the smallest to the largest.

| 4 | 5 | 6 | 7 | 8 | 9 |
|---|---|---|---|---|---|
| □ $<$ | □ $<$ | □ $<$ | □ $<$ | □ $<$ | □ |

7. Order the following numbers from the largest to the smallest.

| 4 | 5 | 6 | 7 | 8 | 9 |
|---|---|---|---|---|---|
| □ $>$ | □ $>$ | □ $>$ | □ $>$ | □ $>$ | □ |

8. Colour the following shapes to make each statement true. The first example has been done for you.

a)  has the same amount as  : 

b)  is more than  : 

c)  is 1 more than  : 

9. Draw lines in the following blank space to make each statement correct.

| | | |
|---|---|---|
| ——— | < | |
| | < | ——— |
| ——— | > | |
| | > | ——— |

10. Compare your height with other classmates and find the tallest person in your class.

In my classroom, __________ is the tallest person.

## Module 3 认识方位 Introduction to positional words

### ❖ 认识上下 Knowing up and down

Some important words for up and down: above, under, on top, below.

Balloons are on top of the cat

A women is under the tree

### ❖ 认识前后 Knowing front and behind

Simon is in front of Lily.
Dan is behind Lily.

### ❖ 认识里外 Knowing inside and outside

Apple is inside the box

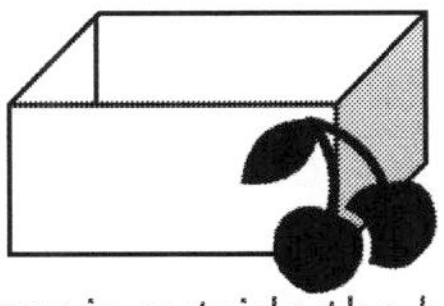

Cherry is outside the box

### ❖ 认识左右 Knowing left and right

Try this: Put your hands in front of your face. Make sure you can see your nails. Open your two thumbs. Which hand looks like a letter "L"?

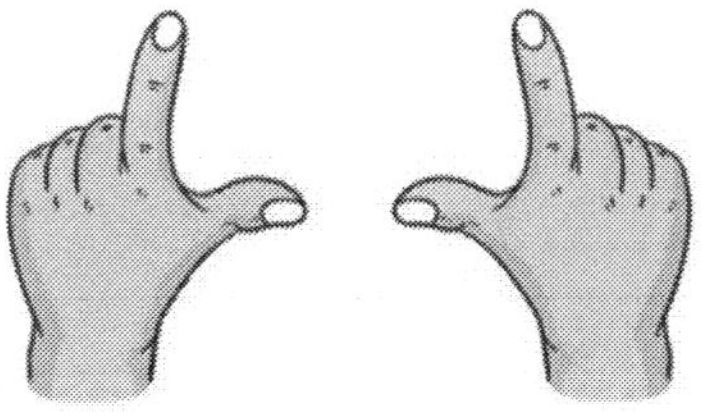

**Module 3 Practice Lesson 1**

11. Circle the correct answer.

a) Coconut is (on / under) the tree.

b) Puppy is (on / under) the tree.

c) Eagle is (above / below) the tree.

12. Decide if the object is 'up' or 'down'.

13. Describe the position of each object in relation to the table.

The toy train is _______ the table

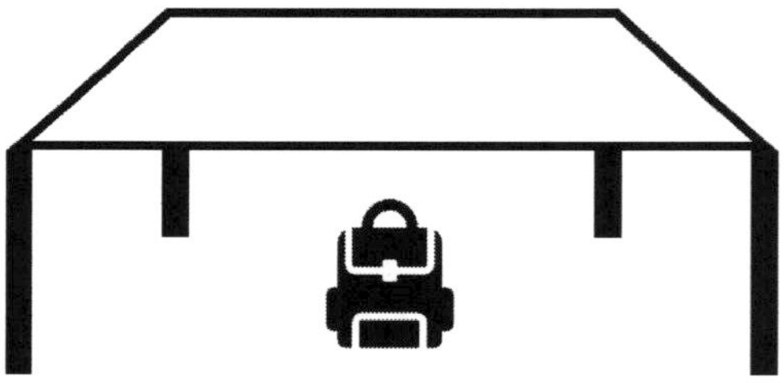

The backpack is _______ the table

14. Describe the position of each person using "above" or "below".

a) Peter is ____________________ Sean.
b) Cici is ____________________ Sean.
c) Sean is ____________________ Peter.
d) Sean is ____________________ Cici.
e) Cici is ____________________ Peter.
f) Peter is ____________________ Cici.

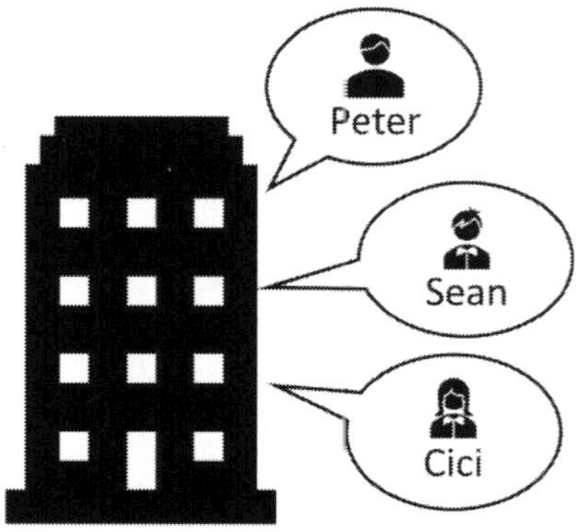

## Module 3 Practice Lesson 2

**12.** Tick ( √ ) the correct graph according to the descriptions provided.

| | |
|---|---|
| The tree that is in the front. 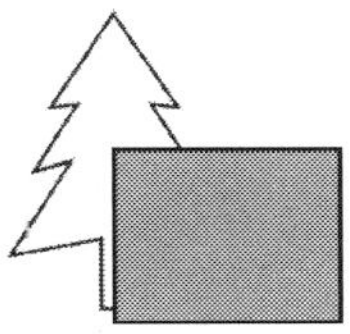 | The bottle that is at the back. 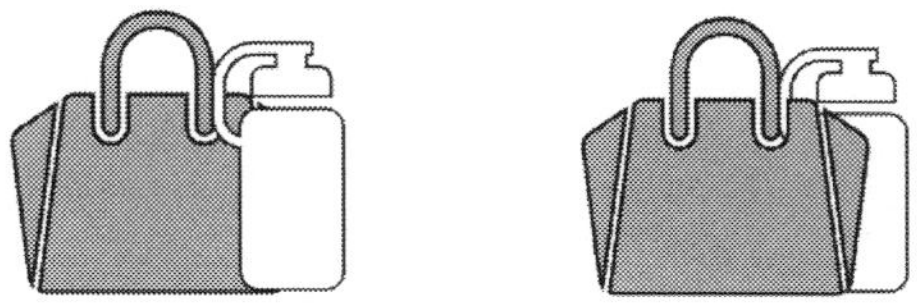 |
| The cat that is in the front. 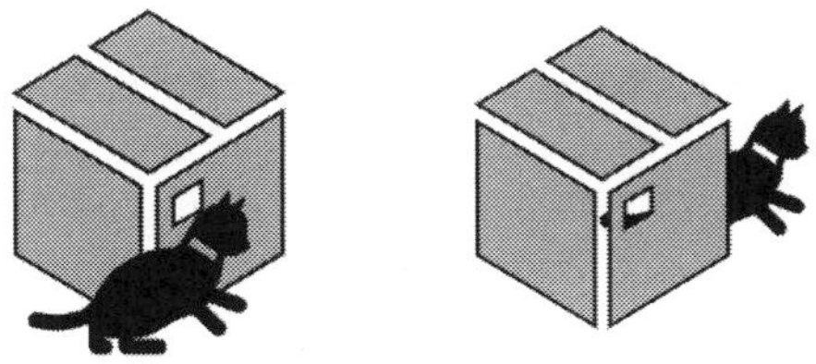 | The apple that is at the back.  |
| The bird that is in the front.  | The pen that is at the back. 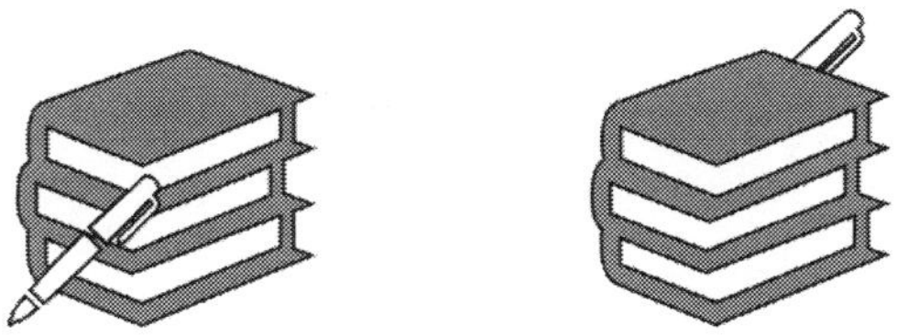 |

**13.** Tick ( √ ) the correct positional word for each statement below.

Cow Fox Goat Horse

a) The fox is ( in front of / at back of ) the goat.

b) The cow is ( in front of / at back of ) the fox.

c) The goat is ( in front of / at back of ) the horse.

d) The fox is in front of ( the goat / the cow ).

e) Looking from the right, the goat is at the ( first / second / third / fourth ) place.

**Module 3 Practice Lesson 3**

11. Circle the correct positional word for each object below (banana/balloon/coat).

| | |
|---|---|
| 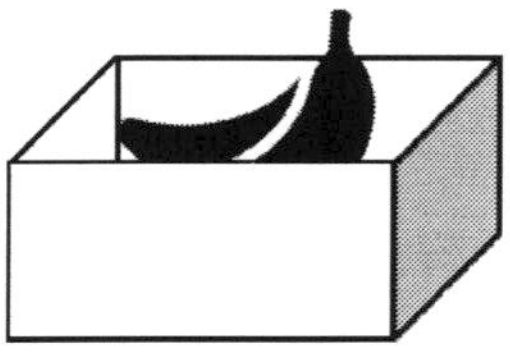 inside outside | 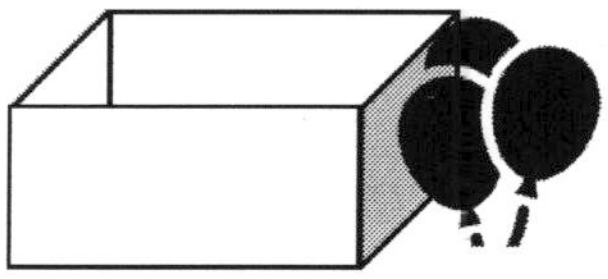 inside outside |
|  inside outside |  inside outside |

12. Write a sentence to describe the object. The first one has been done for you.

The dolphin is outside the water

## Module 3 Practice Lesson 4

11. Colour the correct animal described by each question.

| Colour the elephant on the right | Colour the kangaroo on the left |
|---|---|
|  | 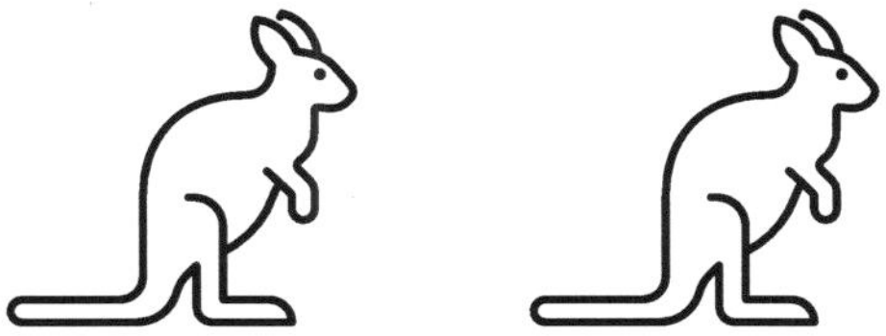 |

| Colour the koala on the left | Colour the fish on the right |
|---|---|
|  | 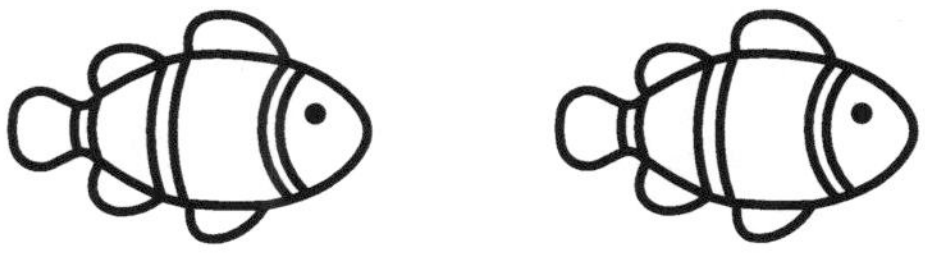 |

12. Circle the correct positional word. The first question has been done for you.

| | |
|---|---|
|  |  |
| The chicken is facing towards ( the right / the left ). | The snail is crawling to the ( left / right ). |

13. Guess the name of each student.

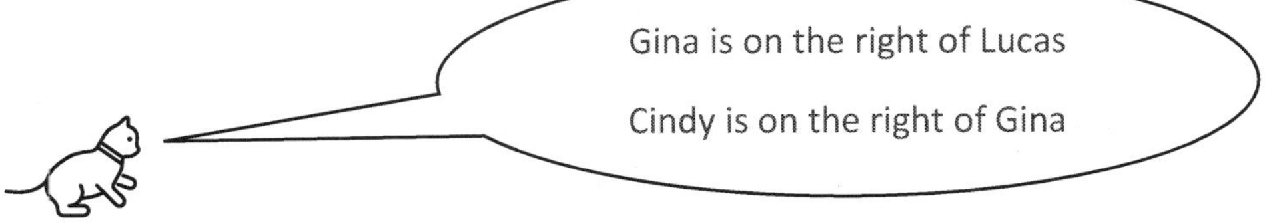

## Module 4 0 – 9 的加减法 Adding and subtracting numbers from 0 – 9

❖ **合与分 Sum and difference**

Every number can be formed by other numbers. For example: 3 apples can be split into 1 apple and 2 apples.

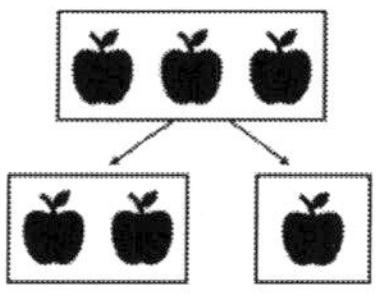

❖ **加法和减法 Addition and subtraction**

➢ **加法 Addition**

Addition is to bring two or more numbers or objects together to make a new total.

The addition sign is ( + ), it means putting things together.

The figure above shows one circle is joining with the other two circles.

2 + 1 = 3 or 1 + 2 = 3

➢ **减法 Subtraction**

Subtraction is to take away one or more numbers or objects.

The subtraction sign is ( – ), it means taking things away.

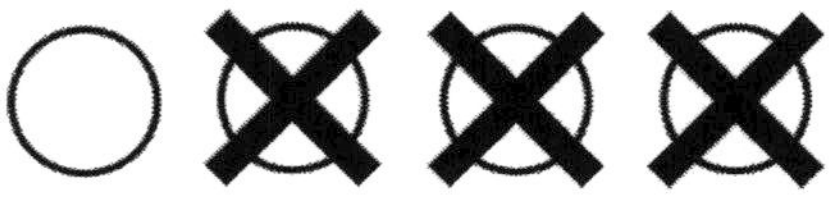

The figure above shows three circles are removed from four circles.

4 – 3 = 1

❖ **0 的认识与加减 Adding and subtracting 0**

0 (zero) means 'nothing'. There are 3 dogs in the dog park, after a while 3 dogs are going back home. How many dogs are in the dog park now?

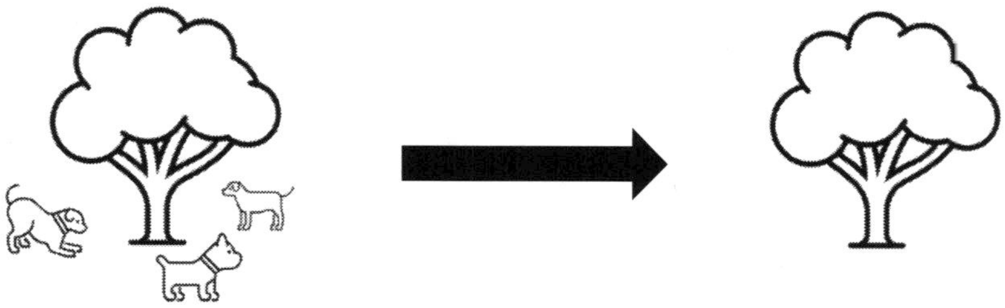

3 – 3 = 0

Think: If you have 5 apples and your teacher gives you another 0 apples, how many apples do you have now?

## Module 4 Practice Lesson 1

15. Fill the boxes with correct number of shapes.

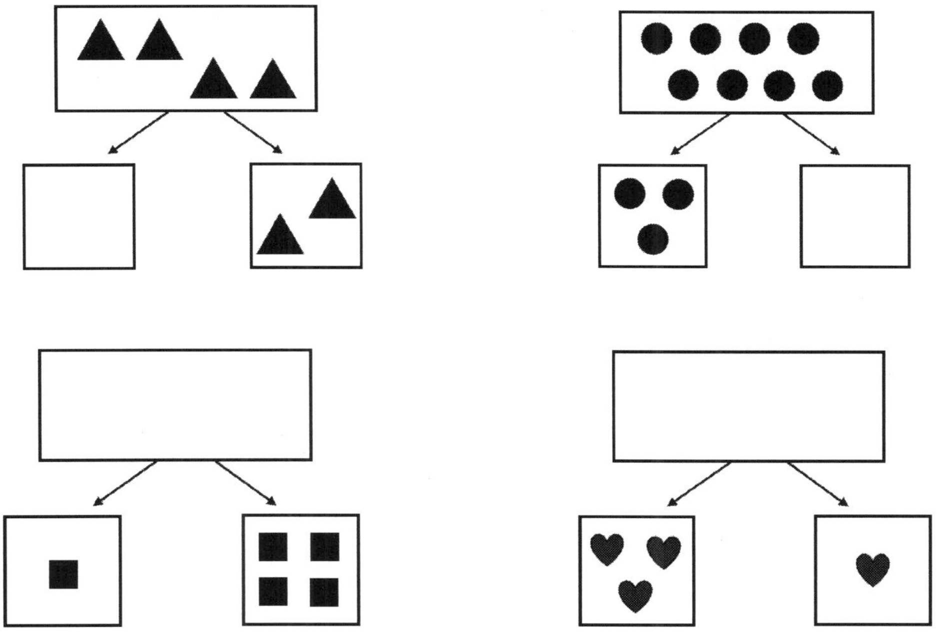

16. Read the diagrams carefully and fill the boxes with the missing values.

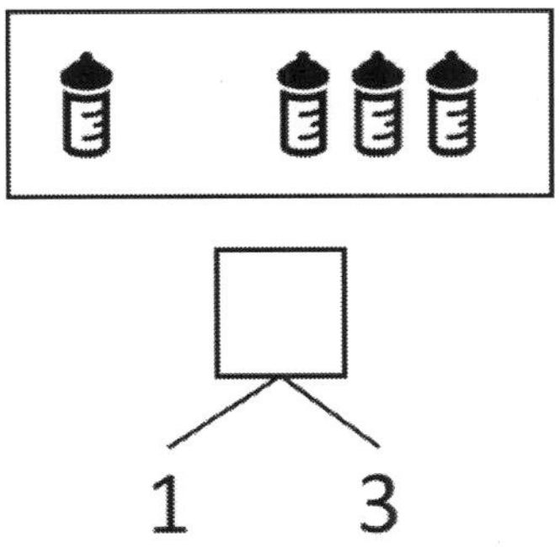


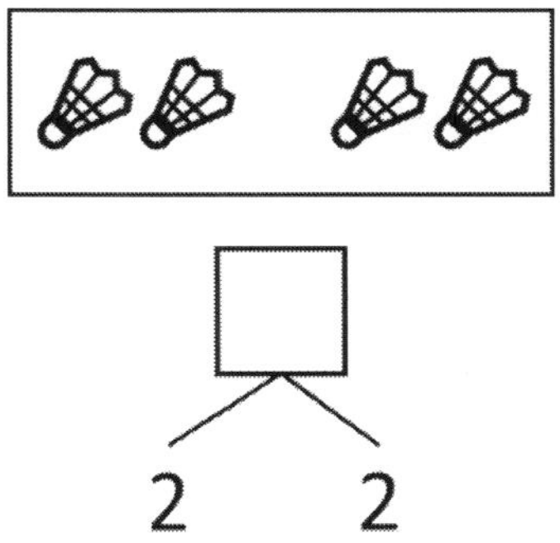


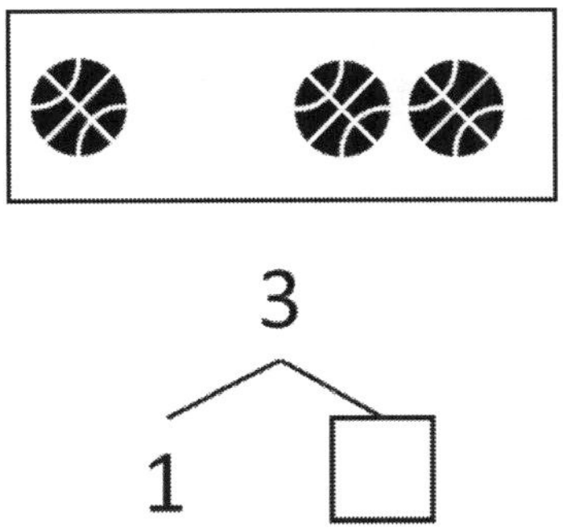


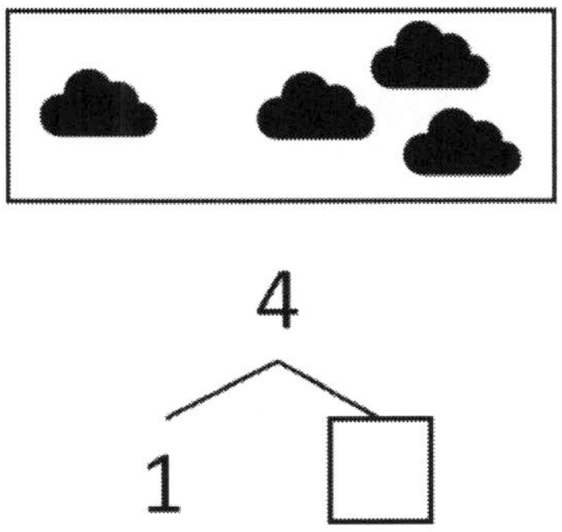

17. Colour the following shapes and write your own answer. The first question has been done for you.

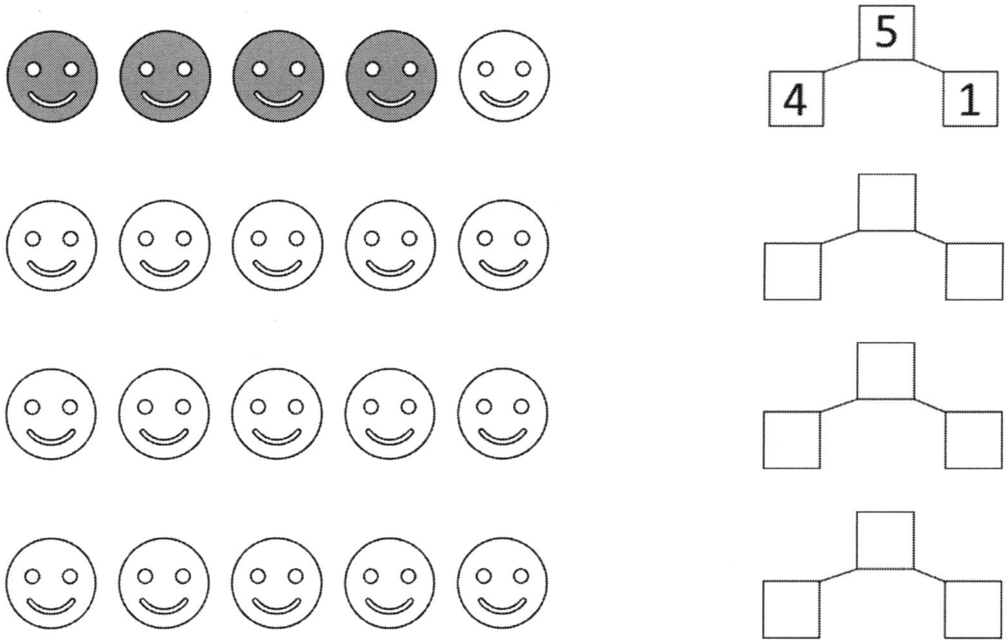


18. Write down the number sentence by looking at the picture. The first question has been done for you.

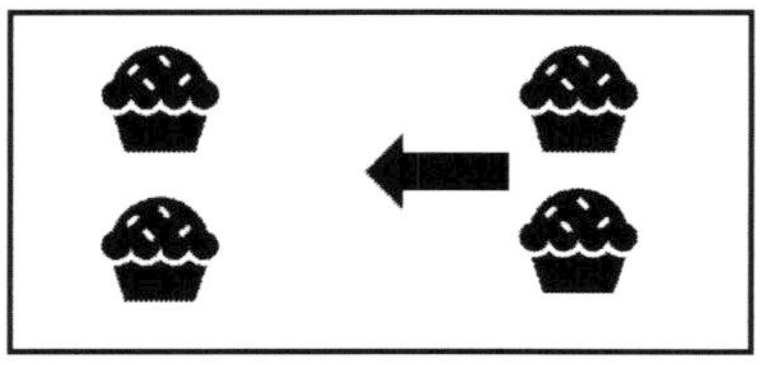

2 + 2 = 4

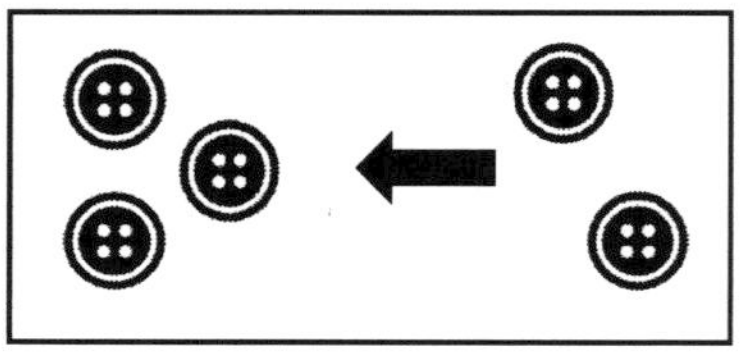

☐ + ☐ = ☐

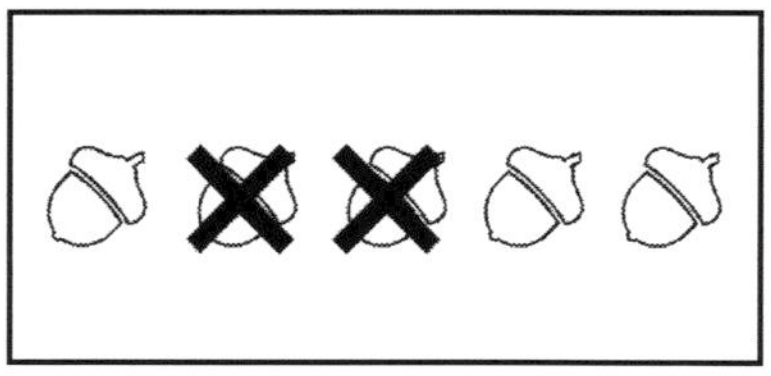

☐ – ☐ = ☐

☐ – ☐ = ☐

## Module 4 Practice Lesson 2

**14.** Draw a picture to describe each number sentence below. You can use any shapes that you like. The first question has been done for you.

| | |
|---|---|
| $3 + 3 = 6$  | $2 + 4 = 6$ |
| $3 + 4 = 7$ | $5 + 2 = 7$ |

**15.** Remove the objects according to each number sentence given. The first question has been done for you.

| | |
|---|---|
| $7 - 3 = 4$  | $8 - 6 = 2$ 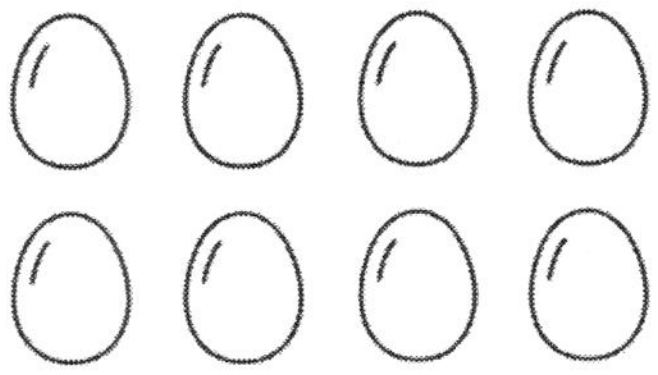 |
| $7 - 6 = 1$ 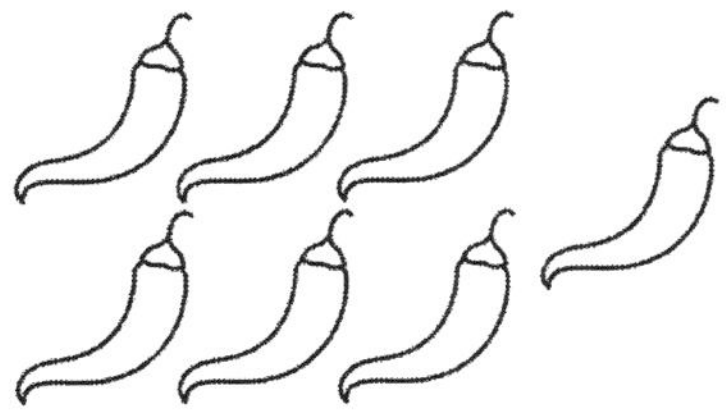 | $9 - 5 = 4$ 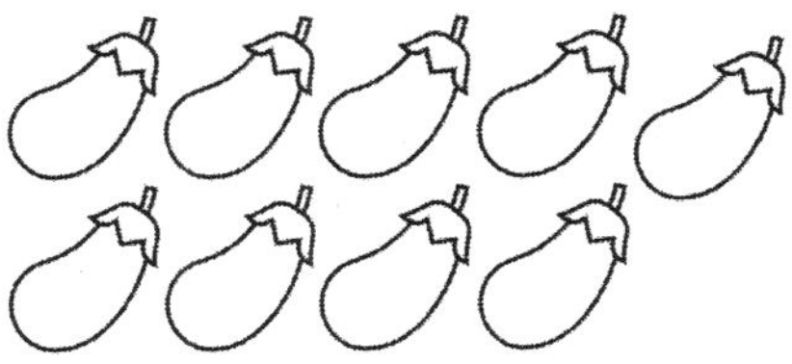 |
| $8 - 5 = 3$ 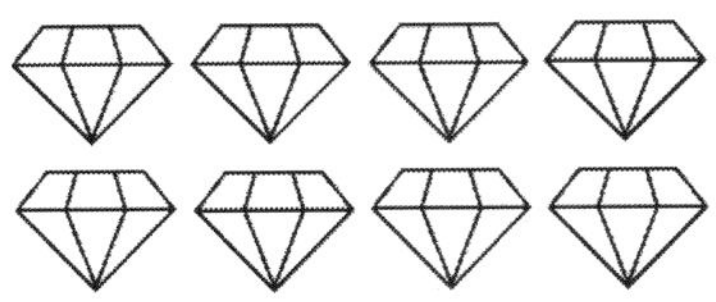 | $4 - 4 = 0$ 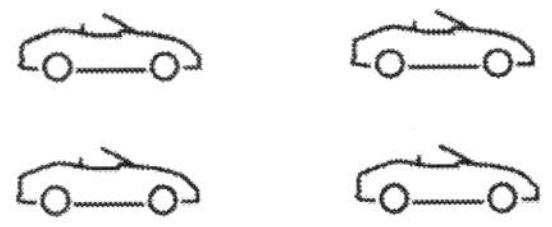 |
| $6 - 5 = 1$  | $6 - 2 = 4$  |

**16.** Use your own way to represent the following number sentences. The first question has been done for you.

| | |
|---|---|
| 3 + 2 = 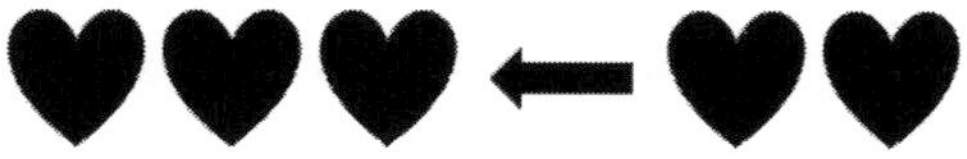 | 4 + 3 = |
| 5 + 1 = | 2 + 7 = |
| 6 − 4 = | 7 − 3 = |
| 3 − 3 = | 4 − 2 = |

**17.** Solve the following questions to get the star.

★

6 − 6 =

3 + 5 = | 7 + 2 =

9 − 3 = | 7 − 5 = | 4 − 2 =

2 + 3 = | 3 + 4 = | 7 + 1 = | 4 + 5 =

18. Look at the examples below and write down your own number sentences in the blank space.

**3** — 1 + 2 = 3 — 4 − 1 = 3 — ☐ — ☐

**6** — ☐ — ☐ — ☐ — ☐

## Module 4 Practice Lesson 3

13. Place numbers into the blank space so that numbers in each line can add up to 9.

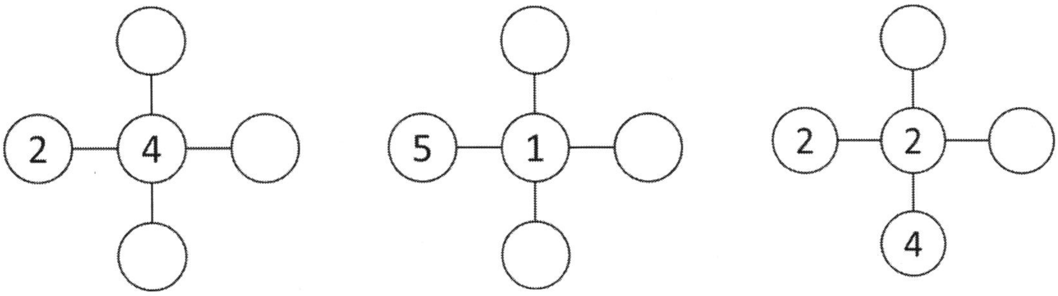


14. Look at the pattern below and fill the missing boxes with correct numbers.

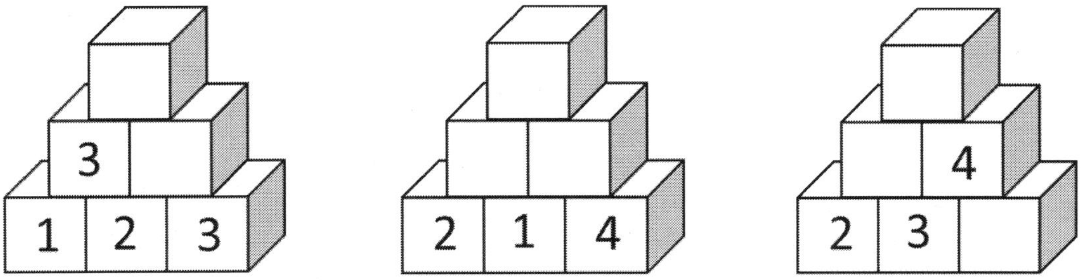


15. Calculate the following number sentences that are related to "0".

$0 + 7 =$ $\quad$ $3 + 0 =$ $\quad$ $4 - 0 =$ $\quad$ $3 - 0 =$

$5 - 5 =$ $\quad$ $7 - 7 =$ $\quad$ $6 - 6 =$ $\quad$ $9 - 9 =$

16. Henry had 8 chocolate bars. He ate 3 bars during the lunch break and ate another 5 bars after dinner. How many chocolate bars was he left with after dinner?

17. Match.

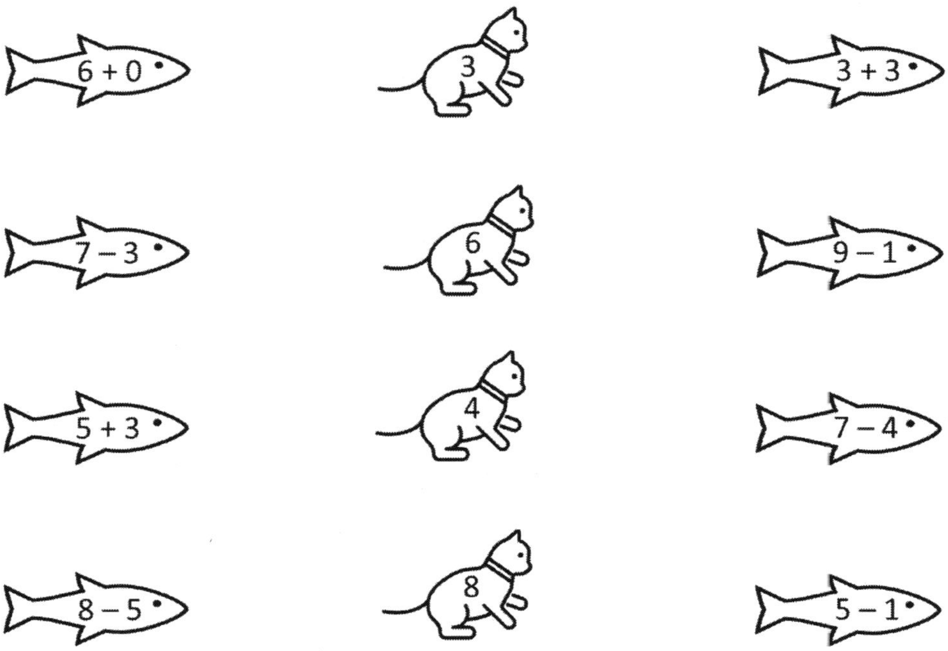


18. Complete each number sentence using number cards listed below. Make sure that each number sentence gives a result of 3.

[2] [6] [4] [1] [7] [0] [5] [3]

[ ] – [ ] = [3]    [ ] – [ ] = [3]

[ ] – [ ] = [3]    [ ] – [ ] = [3]

19. Write the correct answer in the blank space and see what you can find out.

[5] + [2] = [ ]    [2] + [5] = [ ]

[7] – [5] = [ ]    [7] – [2] = [ ]

[4] + [3] = [ ]    [3] + [4] = [ ]

[7] – [3] = [ ]    [7] – [4] = [ ]

## Module 4 Practice Lesson 4

14. Put "+" or "–" sign in the blank space to make each number sentence true.

| | | |
|---|---|---|
| $4 \bigcirc 3 = 7$ | $6 \bigcirc 5 = 1$ | $1 \bigcirc 6 = 7$ |
| $0 \bigcirc 3 = 3$ | $7 \bigcirc 1 = 6$ | $7 \bigcirc 6 = 1$ |
| $2 \bigcirc 5 = 7$ | $3 \bigcirc 1 = 2$ | $6 \bigcirc 2 = 4$ |

15. Gina is trying to connect all of her marbles with a string.

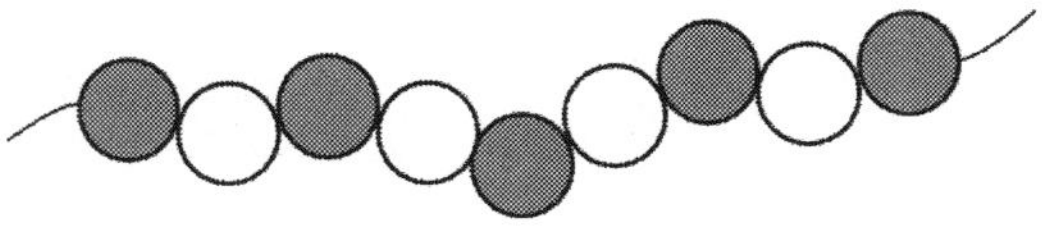

a) How many white marbles does she have?

b) How many black marbles does she have?

c) How many marbles does she have altogether?

16. Fill the missing values.

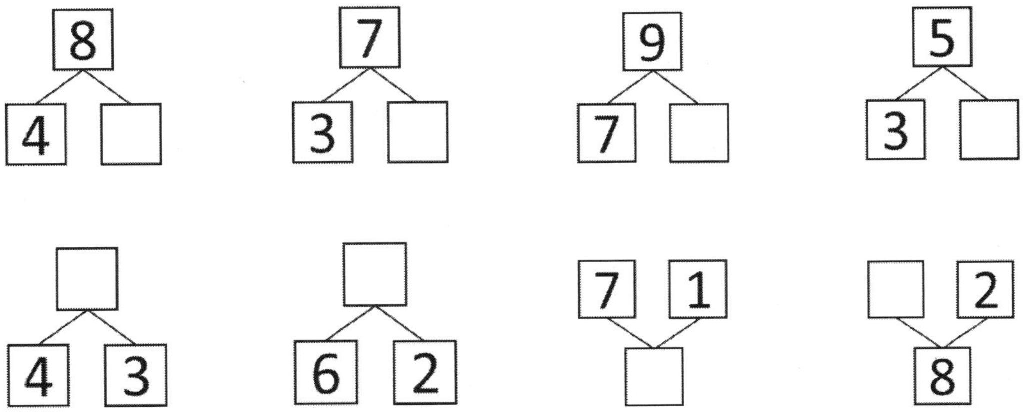

17. Connect the two numbers that can give a sum of 9.

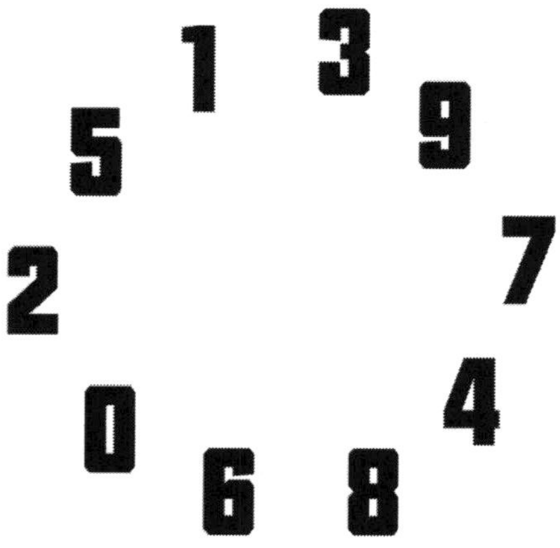


18. Fill in the blank space to make each diagram true.

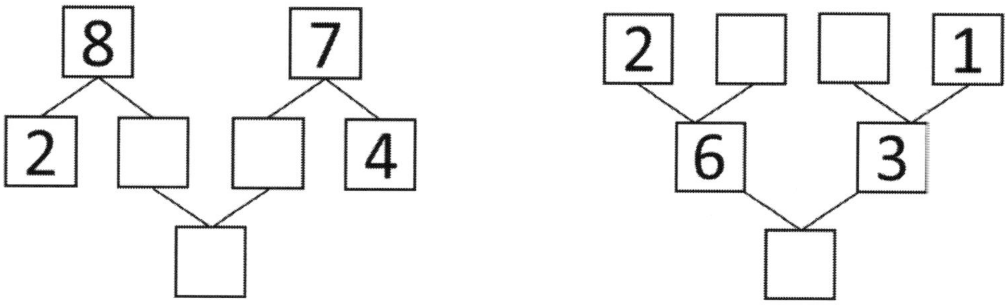


19. Use the following numbers to write as many number sentences as you can.

| | | | | | | | |
|---|---|---|---|---|---|---|---|
| 1 | 7 | 2 | 0 | 4 | 5 | 6 | 3 |

□ + □ = □ □ − □ = □

□ + □ = □ □ − □ = □

□ + □ = □ □ − □ = □

□ + □ = □ □ − □ = □

## Module 5 认识图形 Shapes

❖ **平面图形 2-Dimensional shapes**

2-Dimensional shapes are the shapes on a plane. It is always flat, like an image on paper. Here are some basic 2D shapes we need to be familiar with:

| | | |
|---|---|---|
| Square |  | 4-sided shape |
| Rectangle |  | 4-sided shape |
| Triangle |  | 3-sided shape |
| Circle |  | Rounded shape |

❖ **立体图形 3-Dimensional solids**

3D solids have three dimensions length, width and height. Everything you can hold in the real world can be defined as a solid. The basic solid that you need to know is the cube.

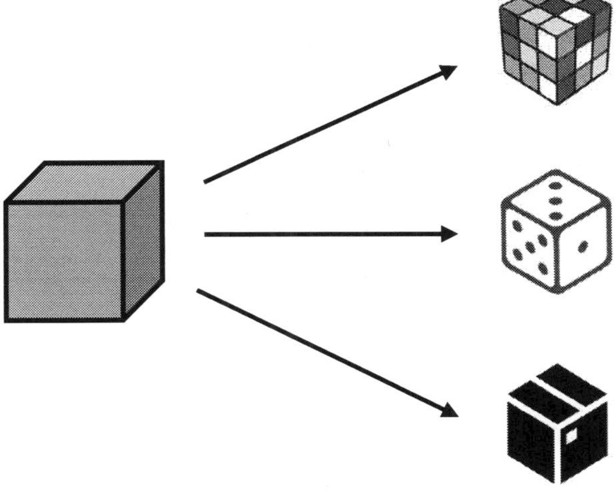

**Module 5 Practice Lesson 1**

19. Connect the shape with correct name.

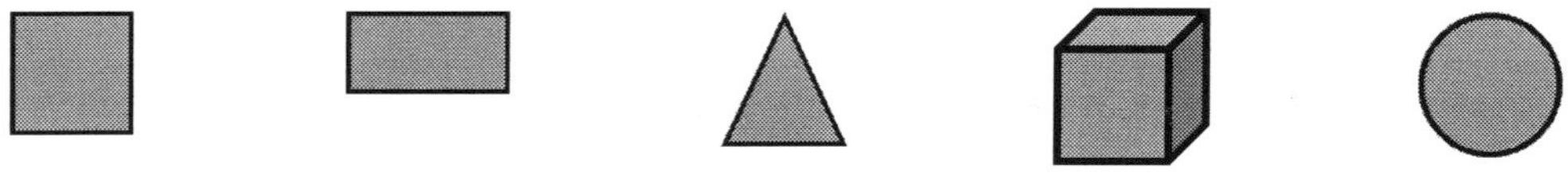

Cube    Circle    Square    Rectangle    Triangle

20. Colour only the same shapes as the one on the left.

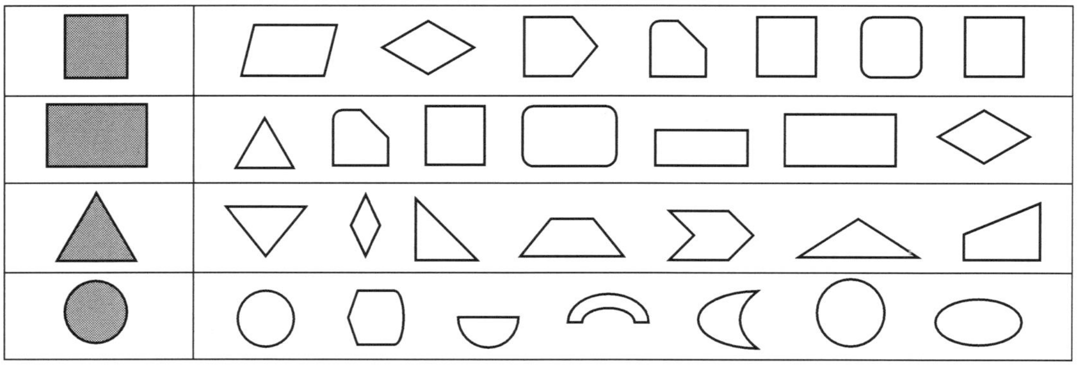

21. Count the number of each shape in the diagram below.

There are __________ circles.

There are __________ squares.

There are __________ rectangles.

There are __________ triangles.

**Module 5 Practice Lesson 2**

**19.** Use a different colour to classify each type of shape.

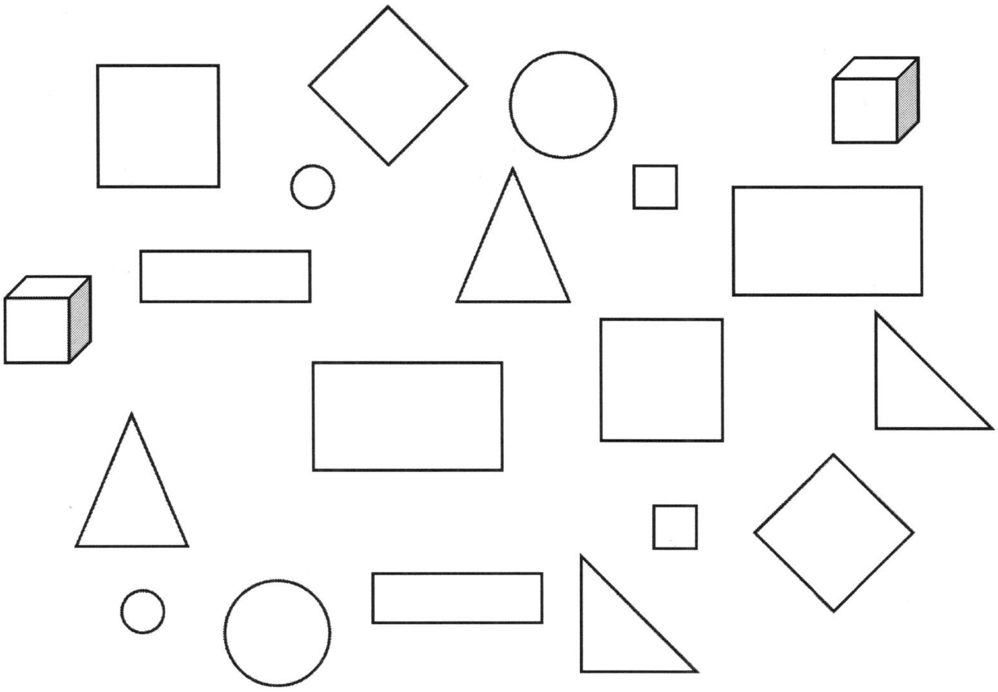

**20.** Draw a line to cut the following shapes. You only can cut each shape once.

| | | |
|---|---|---|
|  |  |  |
| Cut the square into 2 rectangles. | Cut the square into 2 triangles. | Cut the square into only 1 triangle. |

| | | |
|---|---|---|
| 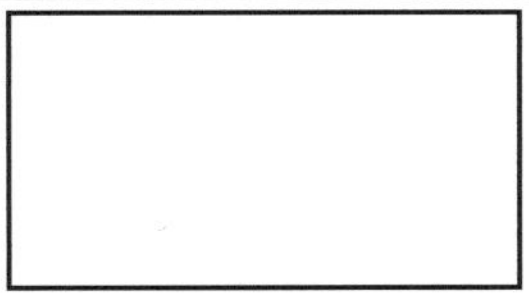 | 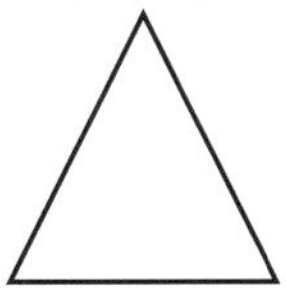 | 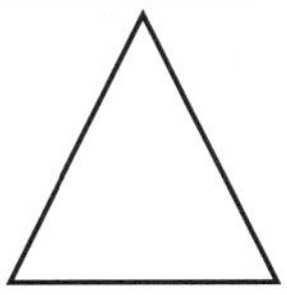 |
| Cut the rectangle into 2 squares. | Cut the triangle into 2 triangles. | Cut the triangle into only 1 triangle. |

## Module 5 Practice Lesson 3

20. Count number of each shape in the following figures.

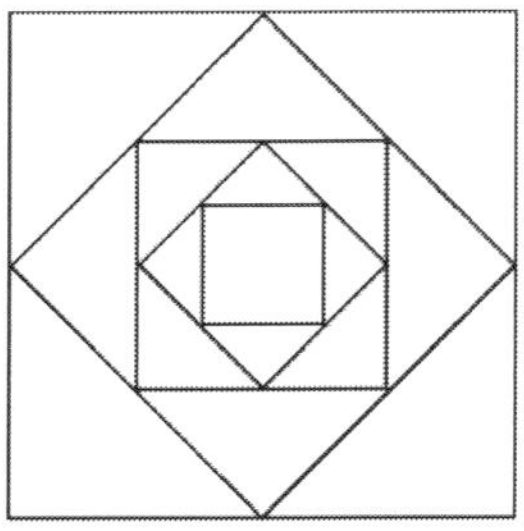

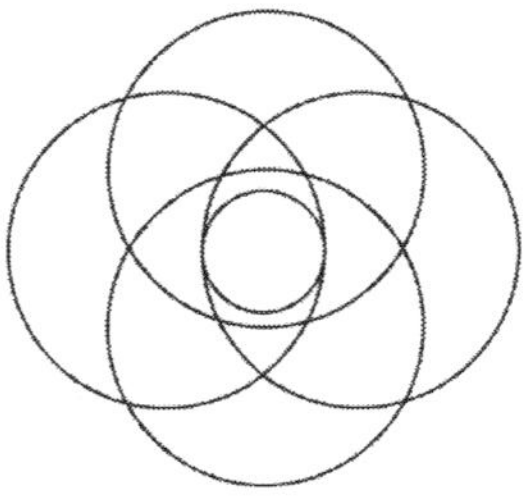

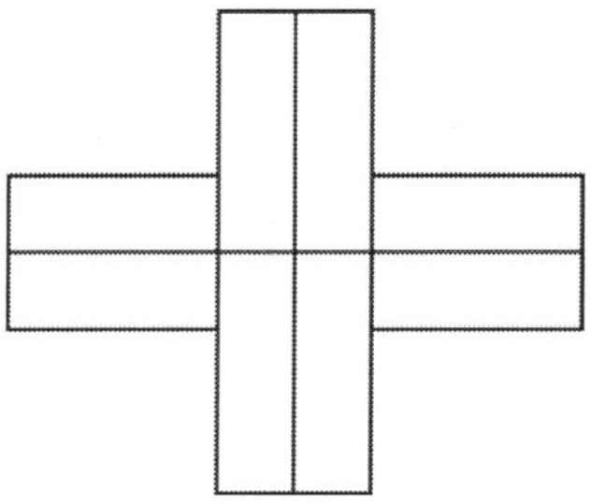

a) How many squares are in the first figure from the <u>left</u>?

b) How many circles are in the second figure from the <u>left</u>?

c) How many rectangles are in the first figure from the <u>right</u>?

21. George wants to fix the wall in his backyard. How many bricks does he need to buy to fix the hole?

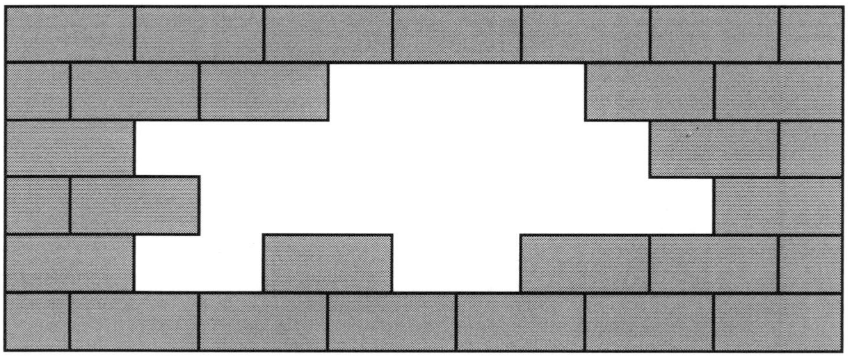

22. A rectangular paper is cut into four of the same sized shape. How many ways can you cut it?

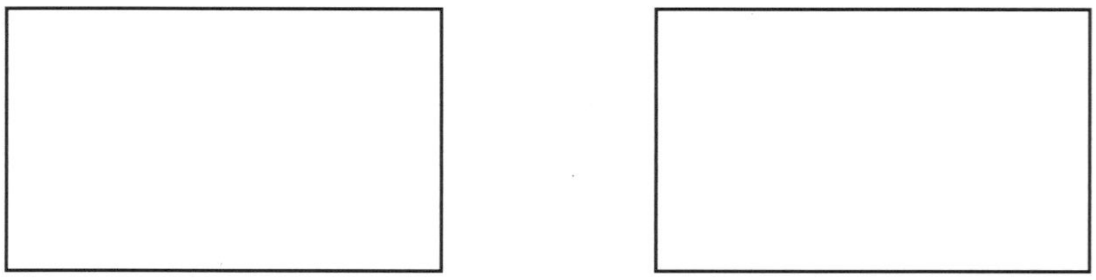

## Module 5 Practice Lesson 4

20. Split the following irregular shapes into regular shapes.

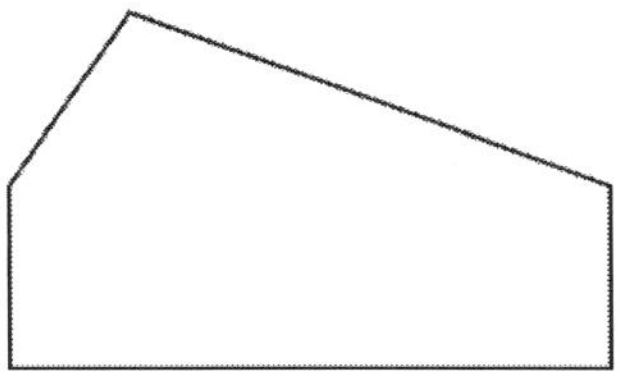

Split the shape into a triangle and a rectangle

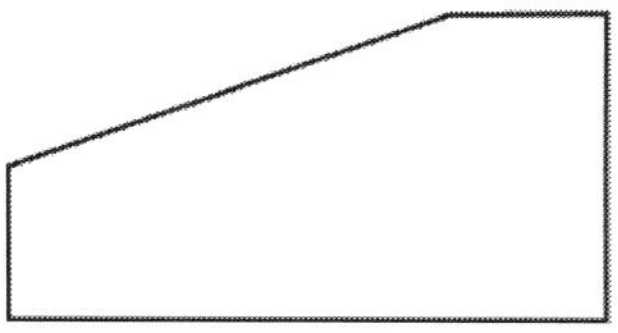

Split the shape into a triangle, rectangle and square

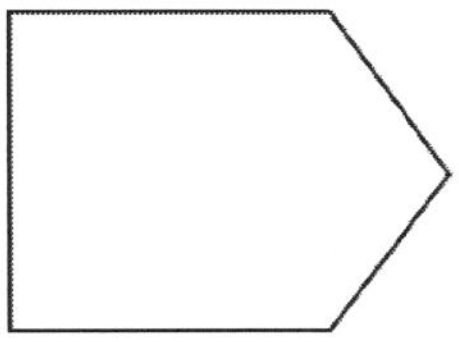

Split the shape into a square and triangle

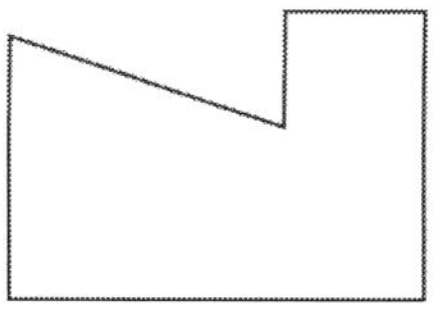

Split the shape into a triangle, rectangle and square

21. Count how many cubes there are in the following figures.

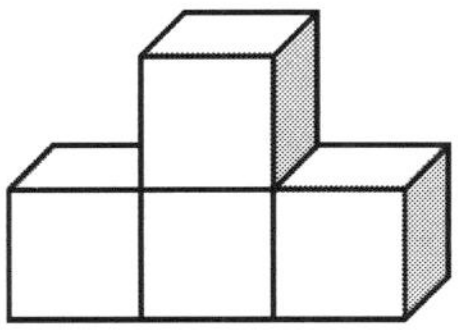

There are __________ cubes

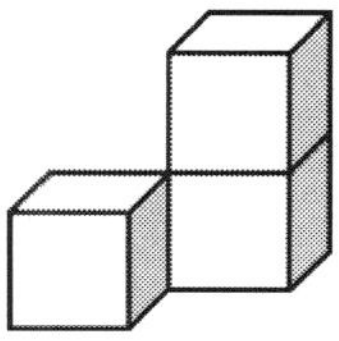

There are __________ cubes

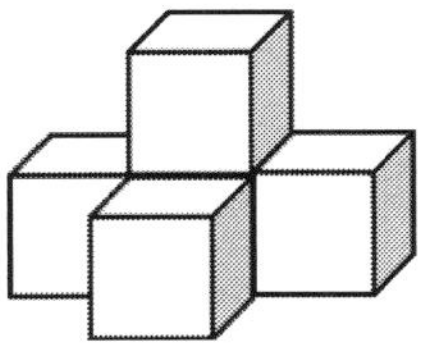

There are __________ cubes

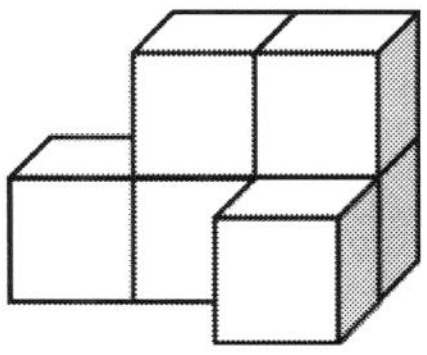

There are __________ cubes

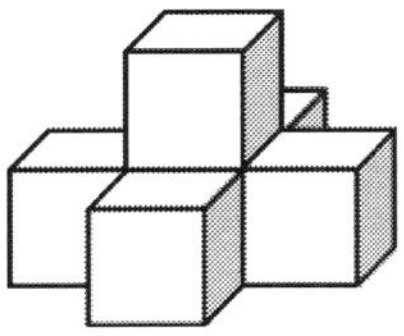

There are __________ cubes

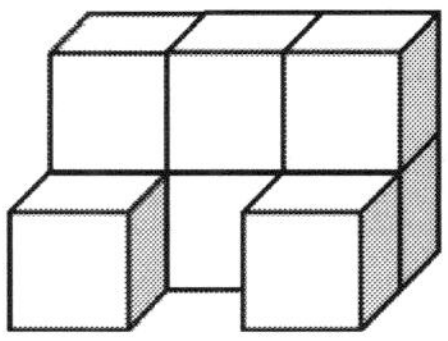

There are __________ cubes

## Module 6 10 – 20 的加减 Adding and subtracting numbers from 10 – 20

❖ **认识 10-20** **Knowing number 10 to 20**

When we introduce number 10 (ten), we need to know the number place first. Each number has their own place. 0 is in the ones place, and 1 is in the tens place.

Imagine there are ten eggs in a basket.

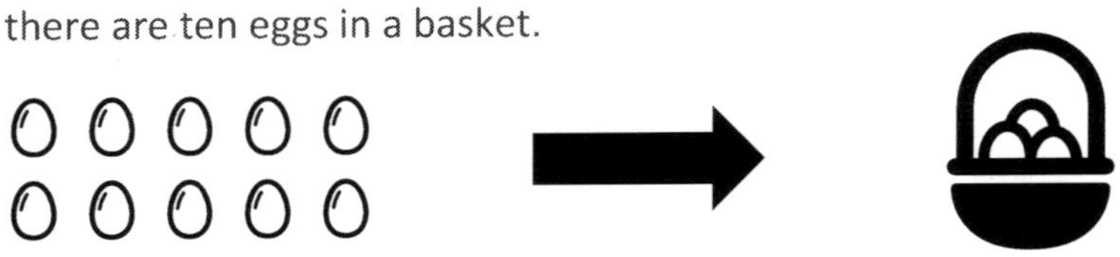

Now let's represent number 10 to 20 below.

| | |
|---|---|
|  | 10 – ten |
|  | 11 – eleven |
|  | 12 – twelve (dozen) |
| 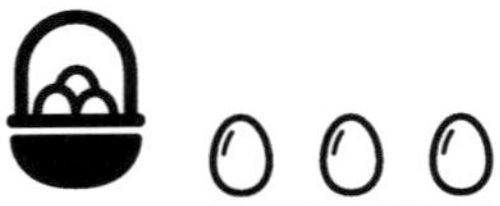 | 13 – thirteen |
|  | 14 – fourteen |
| 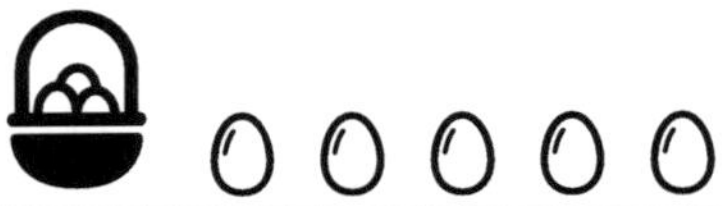 | 15 – fifteen |
|  | 16 – sixteen |
| 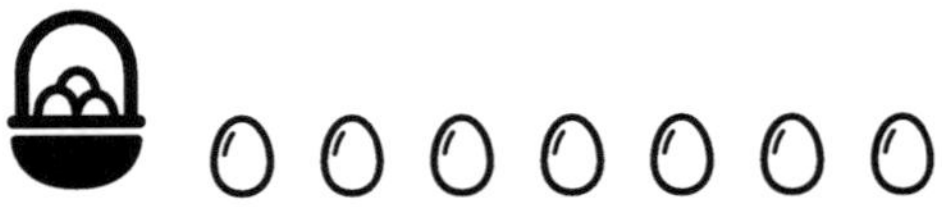 | 17 – seventeen |
|  | 18 – eighteen |
| 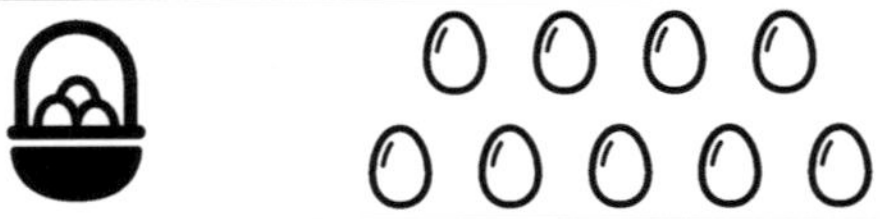 | 19 – nineteen |
| 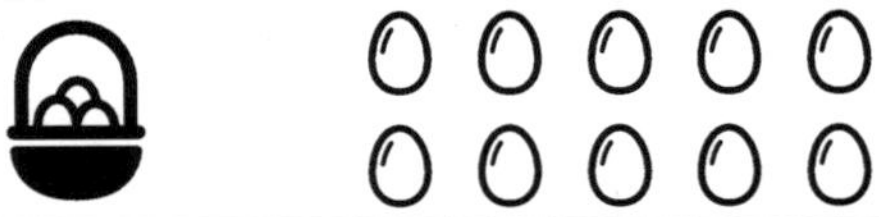 | 20 – twenty |

- **10-20 的加减 addition and subtraction of 10 to 20**

If I have a two-digit number adding with a one-digit number, make sure you add their ones place together.

For example: There are 10 sticks. If I add 1 more stick, there will be 11 sticks.

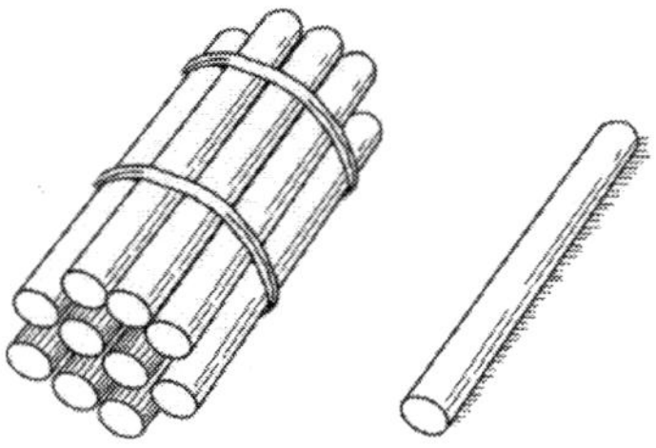

10 + 1 = 11 sticks

Now, if I have 14 sticks. And my friend Jenny takes away 4 sticks. How many do I have now?

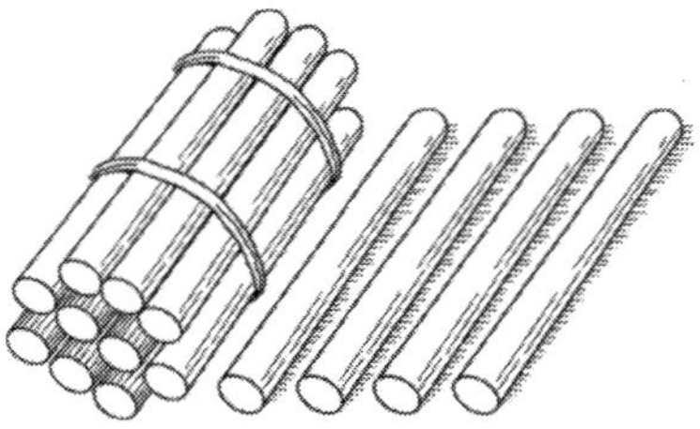

14 – 4 = 10 sticks

- **连加连减 repeated addition and subtraction**

There are 10 sticks. If my friend John gives me 2 sticks, and Jason gives me another 2 sticks. How many sticks do I have now?

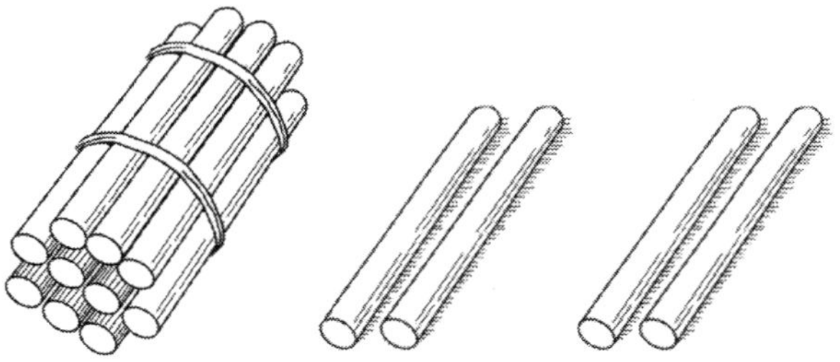

10 + 2 + 2 = 14 sticks

There are 19 sticks. If my friend John takes 3 sticks away, and Jason takes another 3 sticks away. How many sticks do I have now?

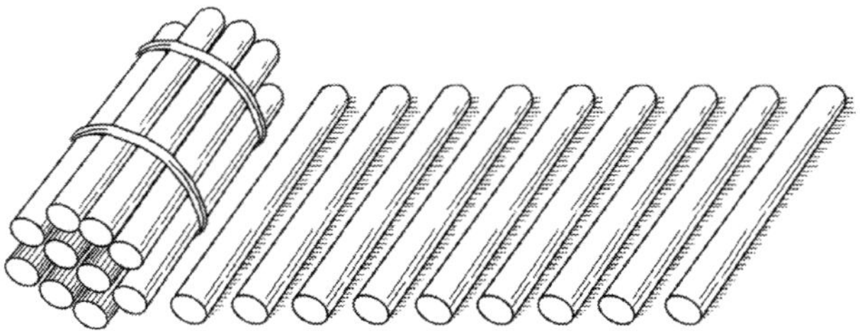

19 – 3 – 3 = 13 sticks

## Module 6 Practice Lesson 1

1. Circle the amount of 10 for each item below.

2. Write the following numbers in words.

**11:**

**15:**

**18:**

**20:**

3. Connect these points in order.

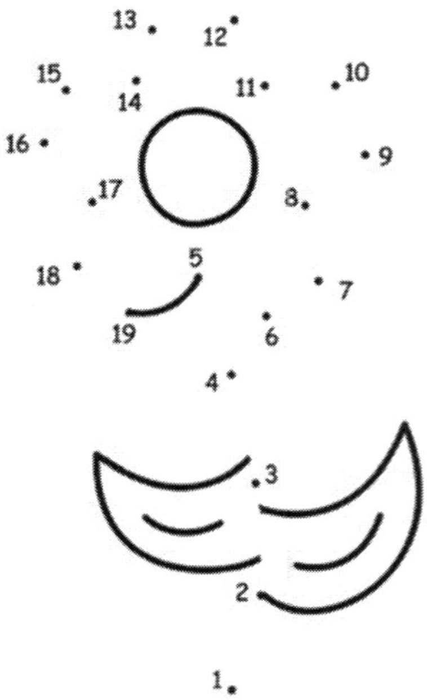

4. Count the amount.

| | |
|---|---|
|  There are ____________ bananas |  There are ______________ fingers |
|  There are ____________ books |  There are ____________ cherries |

5. What number is showing on the abacus?

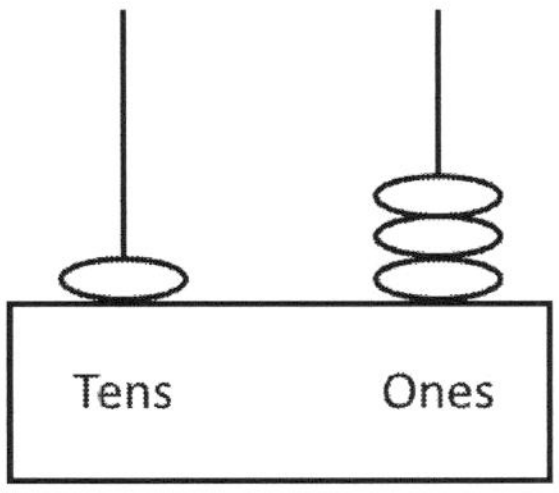


__________________

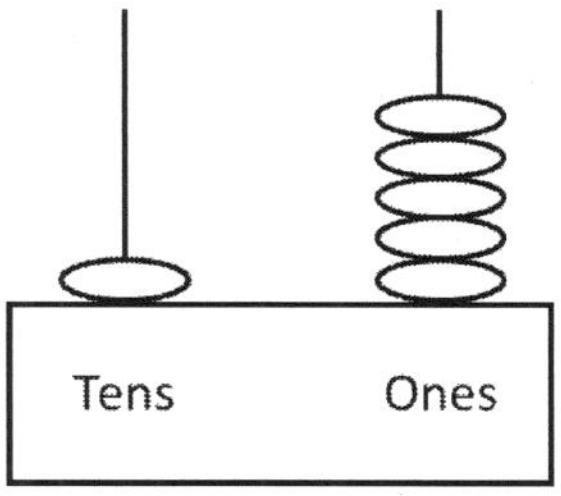


__________________

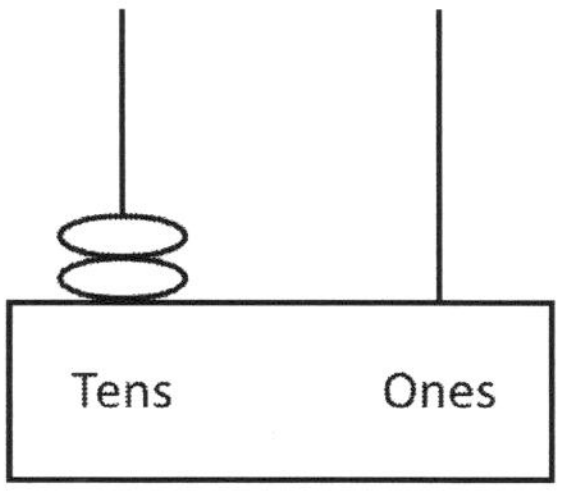


__________________

6. Find the pattern and fill in the missing blanks.

| | | | | | | | | | |
|---|---|---|---|---|---|---|---|---|---|
| 11 | | 13 | | | 16 | | | | |

| | | | | | | | | | |
|---|---|---|---|---|---|---|---|---|---|
| 20 | | | 17 | | | 14 | | | 11 |

## Module 6 Practice Lesson 2

1. Draw the beads according to the number shown below.

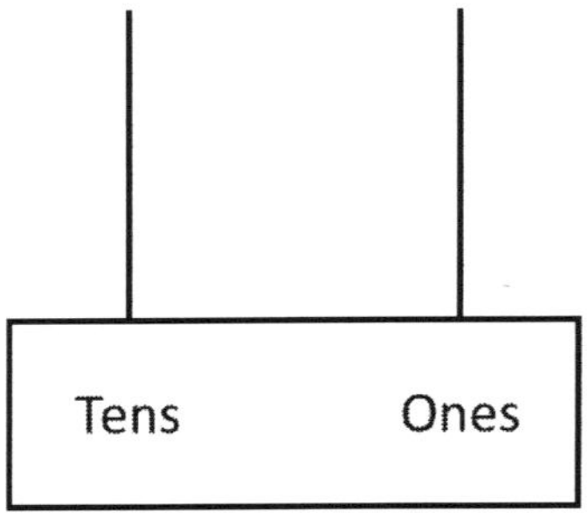


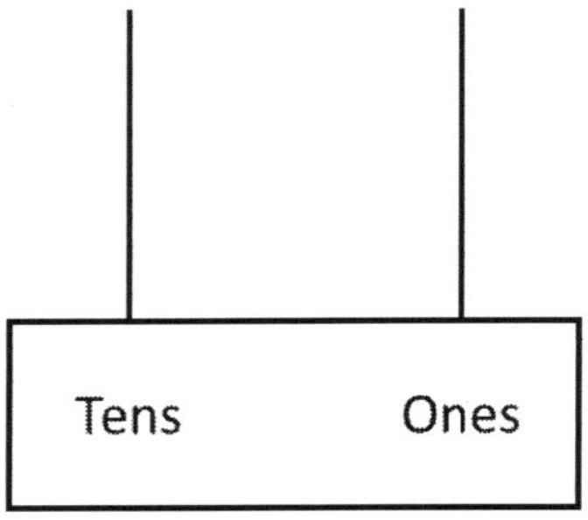


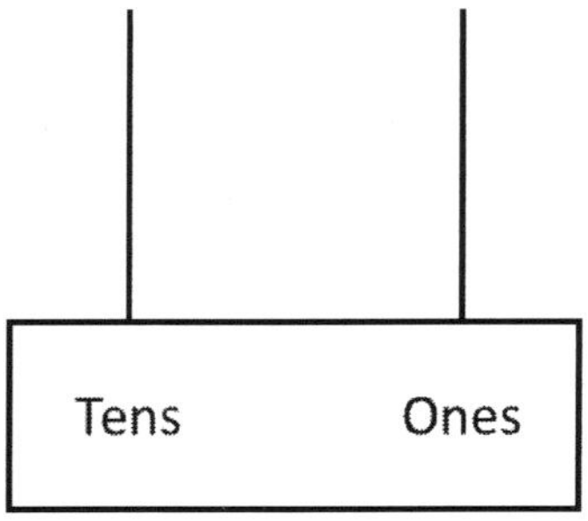


14 12 17

2. Write the correct number to make the number sentence true. The first question has been done for you.

| | |
|---|---|
| 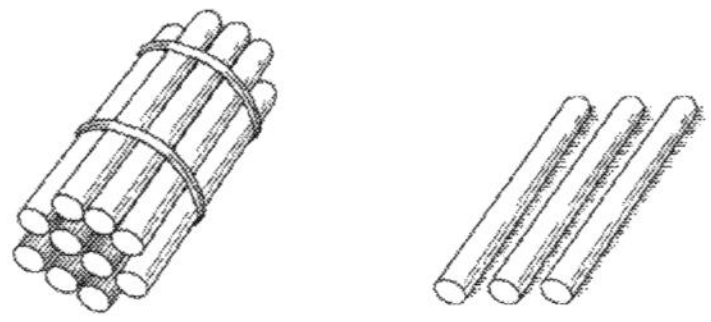 | 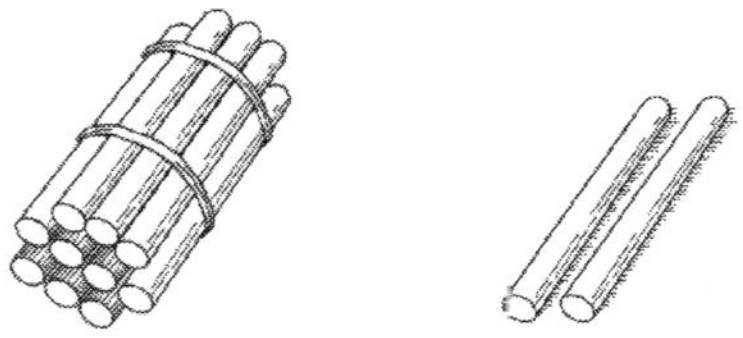 |
| 10 + 3 = 13<br>3 + 10 = 13<br>13 − 10 = 3<br>13 − 3 = 10 | □ + □ = □<br>□ + □ = □<br>□ − □ = □<br>□ − □ = □ |
| 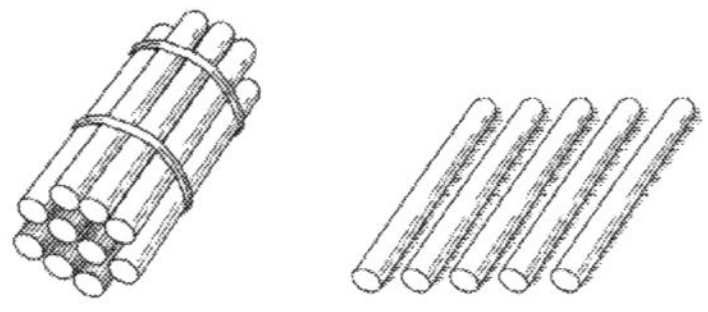 | 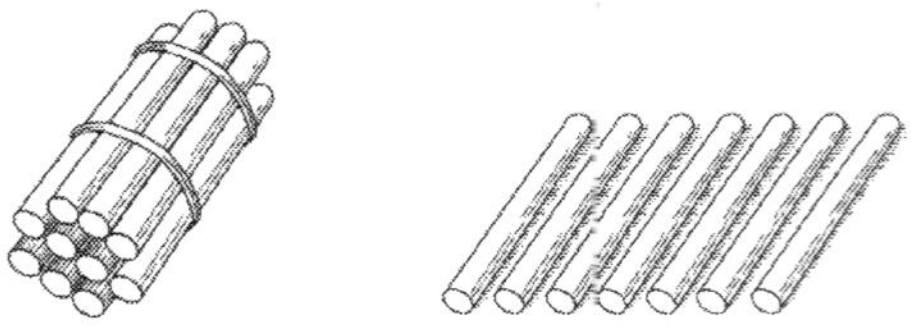 |
| □ + □ = □<br>□ + □ = □<br>□ − □ = □<br>□ − □ = □ | □ + □ = □<br>□ + □ = □<br>□ − □ = □<br>□ − □ = □ |

3. Write down the correct number sentence according to the figure.

? chicken

14 chicken

□ ○ □ = □

? volleyballs

19 volleyballs

□ ○ □ = □

? apples

14 apples

□ ○ □ = □

? people

15 people

□ ○ □ = □

4. Connect the same pair of apples.

18 – 8

17 – 4

12 + 3

18 – 2

13 + 3

19 – 4

20 – 10

12 + 1

## Module 6 Practice Lesson 3

1. Complete the following table to find the sum and difference.

| | | | | | | |
|---|---|---|---|---|---|---|
| ○ | 10 | 10 | 14 | 17 | 8 | 10 |
| △ | 5 | 2 | 3 | 2 | 3 | 9 |
| Sum | | | | | | |

| | | | | | | |
|---|---|---|---|---|---|---|
| ○ | 17 | 10 | 15 | 18 | 14 | 19 |
| △ | 3 | 2 | 3 | 7 | 2 | 8 |
| Difference | | | | | | |

2. Compare the numbers with ">", "<" or "=".

17 + 2 ________ 18     10 + 5 ________ 19     14 + 3 ________ 17

16 – 5 ________ 11     18 – 6 ________ 10     17 – 4 ________ 15

14 + 0 ________ 15     13 – 3 ________ 9     9 + 2 ________ 10

3. School athletics competition to be held today was delayed by 3 days due to raining. When will the competition reopen again?

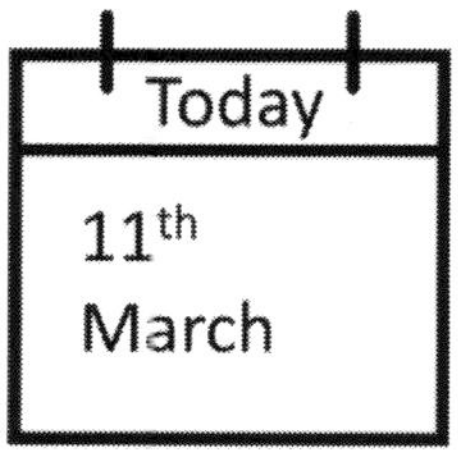


4. Jenny has a story book. She read 6 pages already and left with 12 pages. How many pages does the story book have in total?

5. There are 18 butterflies in the garden. If 7 of them flew away, how many butterflies are in the garden now?

6. I am a two-digit number. My ones place is 5 and my tens place is 1. What number am I?

I am ______________________________

7. Olivia had 17 marbles. She gave her friend Lily 2 marbles and lost 3 marbles. How many marbles does Olivia have now?

8. There are 15 passengers in the bus. At one stop, 3 passengers got off the bus and 5 passengers got on the bus. How many passengers are on the bus now?

## Module 6 Practice Lesson 4

1. Find the missing numbers in each grid. The sum of the numbers in each row and column are given.

| | | |
|---|---|---|
| 4 | | 10 |
| 6 | 5 | 11 |
| 10 | 11 | |

| | | |
|---|---|---|
| 3 | | 10 |
| 6 | 10 | 16 |
| 9 | 17 | |

| | | |
|---|---|---|
| 5 | | 10 |
| 9 | | 11 |
| 14 | 7 | |

| | | |
|---|---|---|
| 11 | | 15 |
| 8 | 10 | 18 |
| 19 | 14 | |

| | | |
|---|---|---|
| | | 20 |
| 4 | 6 | 10 |
| 14 | 16 | |

| | | |
|---|---|---|
| | 7 | 18 |
| 4 | | 14 |
| 15 | 17 | |

2. What number did I start with?

Start ☐ → Add 7 → 17

Start ☐ → Add 5 → 19

Start ☐ → Minus 5 → 11

Start ☐ → Minus 3 → 15

3. Yeh has $15 in his wallet. He buys a $5 pencil and $1 rubber. How much money he will left with?

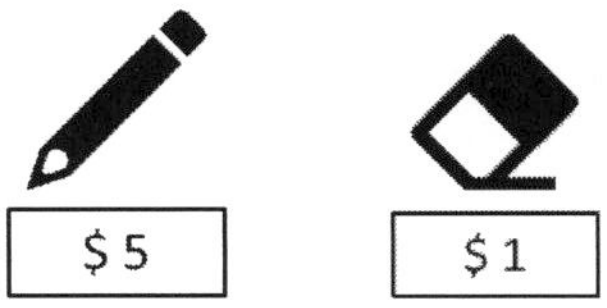


4. Cecilia need to remember 18 Chinese characters every day. Today she already put 7 words in her mind. How many characters she still need to remember?

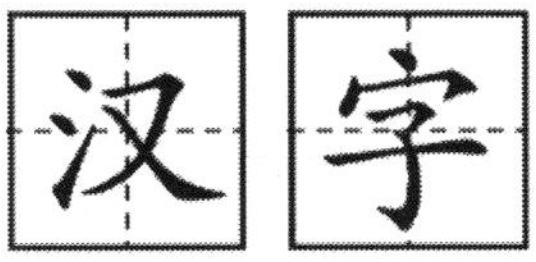


5. There are 10 cars in the carpark, and 5 more cars came after a while. 4 cars were left afterwards. How many cars are in the car park now?

6. There are 20 students in the classroom. 14 of them are boys.

a) how many girls are in the classroom?

b) If 5 students left the classroom, how many students are in the room now?

## Module 7 认识钟表 Introduction to time

Time is an invisible and intangible thing. But it is very important in our daily lives.
Our ancestors knew the existence of time long ago, and they developed what we know as the current time by observing the movement of the sun.
Nowadays, we have clocks to tell us what the time is.

### ❖ 认识时针和分针 Knowing the hour and minute hands

Clocks are made from four parts. The clock face, hour hand, minute hand and the second hand.

- The shorter hand is called the **hour hand**, it shows the hours.

2 hours

4 hours

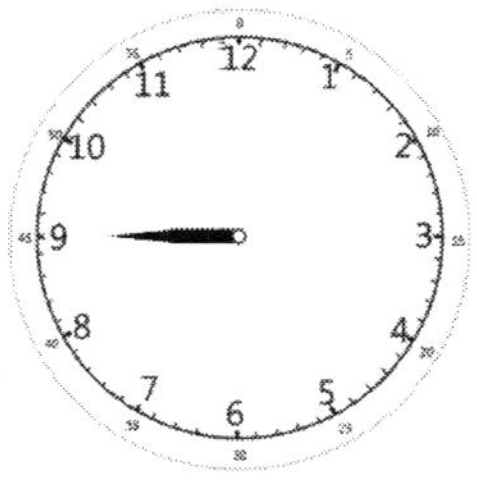

9 hours

- The longer hand is called the **minute hand**, it shows the minutes.

O'clock

15 minutes or Quarter-Past

30 minutes or Half-Past

### ❖ 整点的认识 Knowing o'clock

Now, we know what the hour and minute hands look like. Let's try to read some easy times.

Hour hand pointing to 1
Minute hand pointing to 0
12:00
Twelve o'clock

Hour hand pointing to 5
Minute hand point to 0
5:00
Five o'clock

Hour hand pointing to 9
Minute hand point to 0
9:00
Nine o'clock

- 半点的认识 **Knowing half past**

Hour hand is pointing in between 1 and 2, but has not reached 2. Therefore, it is still in the 1 hour.
Minute hand is pointing to 30

1:30
or
Half-Past One

Hour hand is pointing in between 4 and 5, but has not reached 5. Therefore, it is still in the 4 hour.
Minute hand is pointing to 30

4:30
or
Half-Past Four

Hour hand is pointing in between 10 and 11, but has not reached 11. Therefore, it is still in the 10 hour.
Minute hand is pointing to 30

10:30
or
Half-Past Ten

- 日程安排 **Time schedules**

We need to do lots of things every day, such as getting out of bed, brushing our teeth, having breakfast, going to school, etc.

We need to know what to do at what time, which requires a schedule.

For example, we will get up at 7:00 and go to school at 8:00.

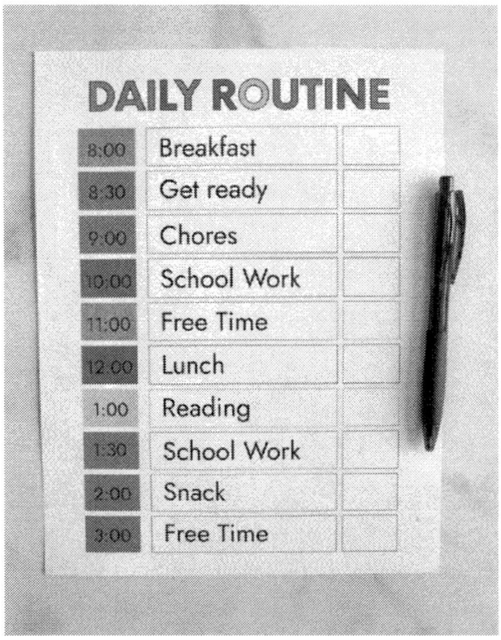

## Module 7 Practice Lesson 1

7. Connect the clock faces with the correct hour time.

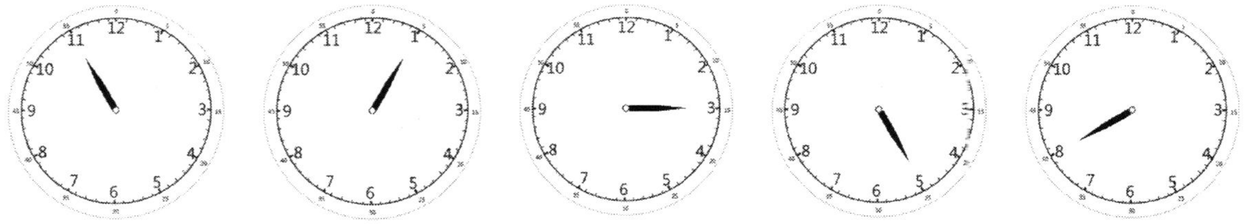

5 hours  8 hours  1 hour  11 hours  3 hours

8. Connect the clock faces with the correct minute time.

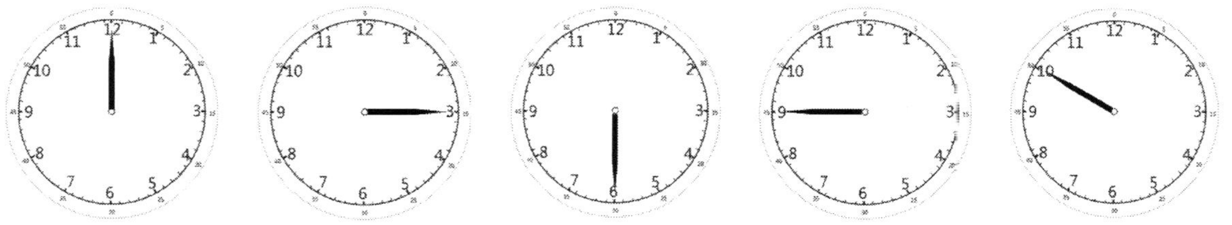

50 minutes  45 minutes  0 minute O'clock  15 minutes Quarter-Past  30 minutes Half-Past

9. Draw the hour hands to show the correct hours.

| | | |
|---|---|---|
| 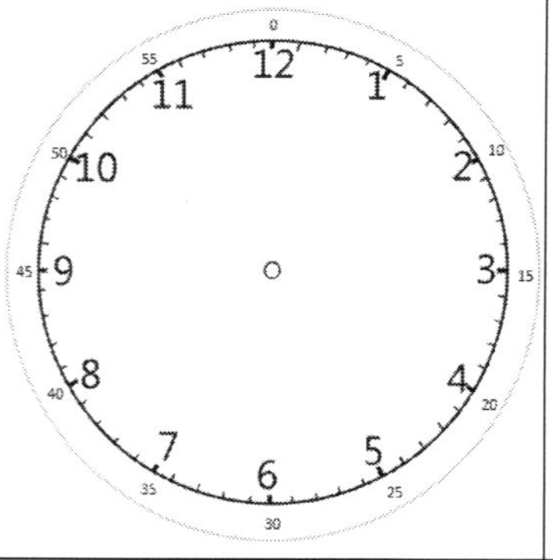 | 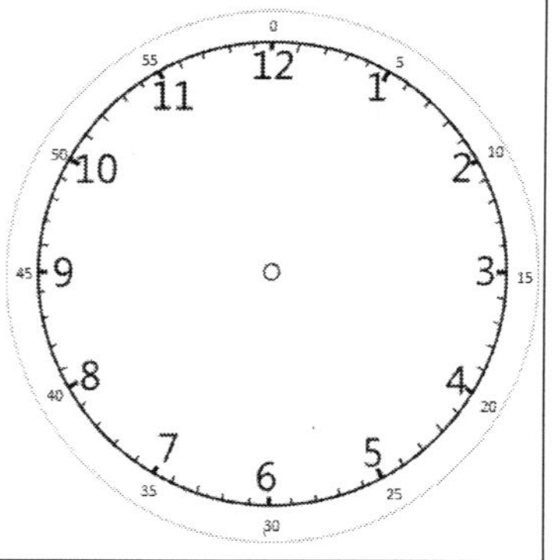 | 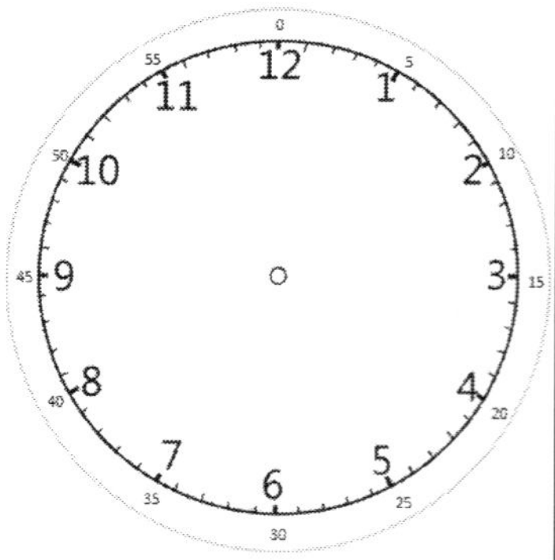 |
| 4 hours | 7 hours | 12 hours |

10. Draw the minute hands to show the correct minutes.

| | | |
|---|---|---|
| 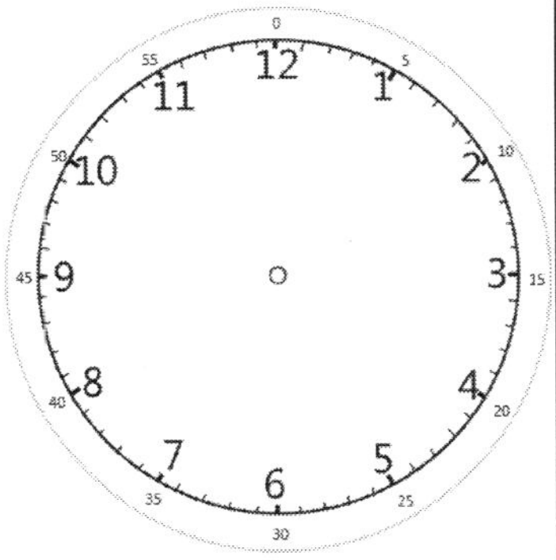 | 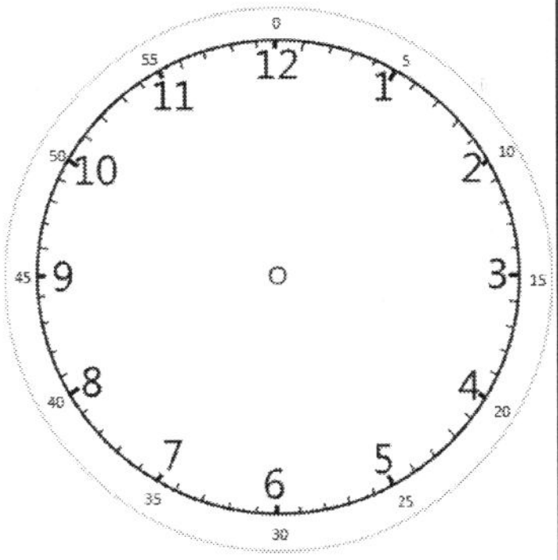 | 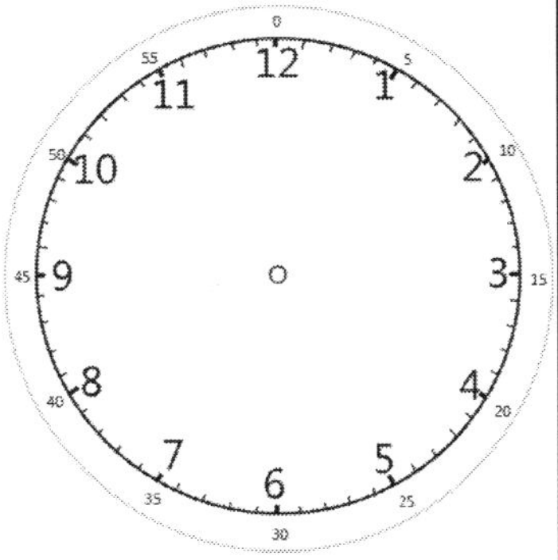 |
| 5 minutes | 20 minutes | 30 minutes |

| | | |
|---|---|---|
| 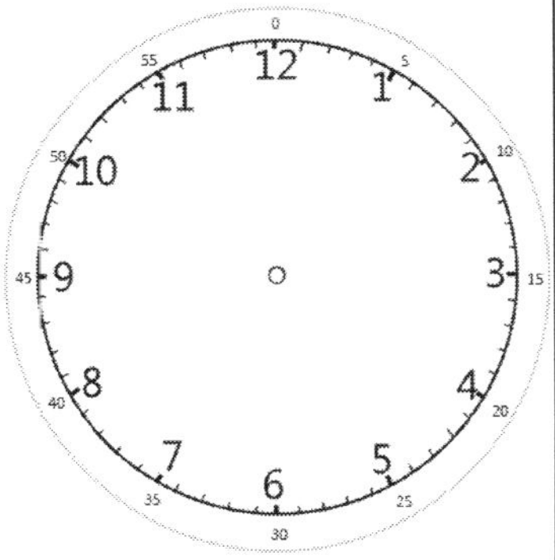 | 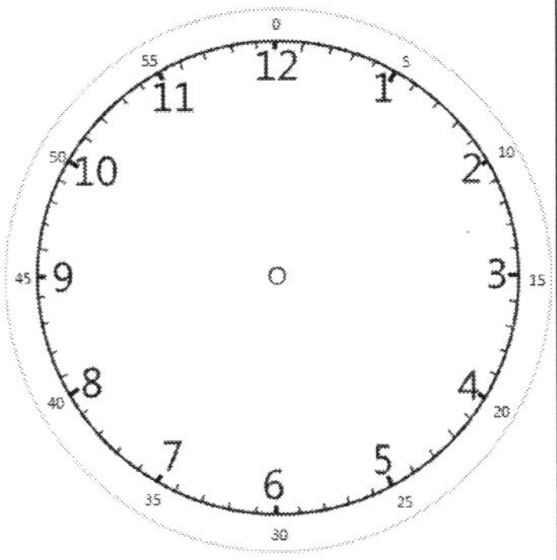 | 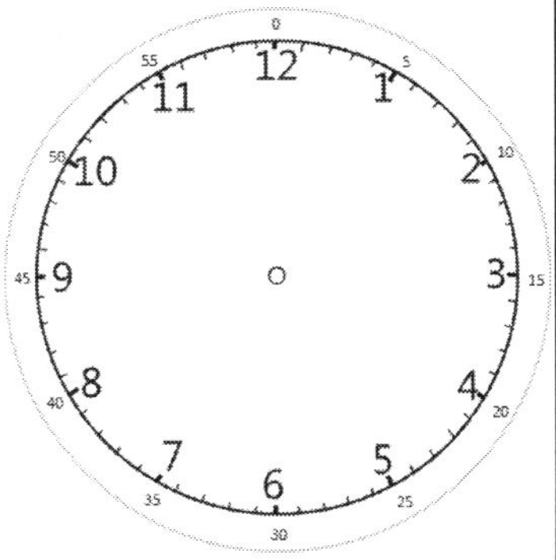 |
| 35 minutes | 45 minutes | 0 minute |

11. Write down the correct hours for the following clock faces.

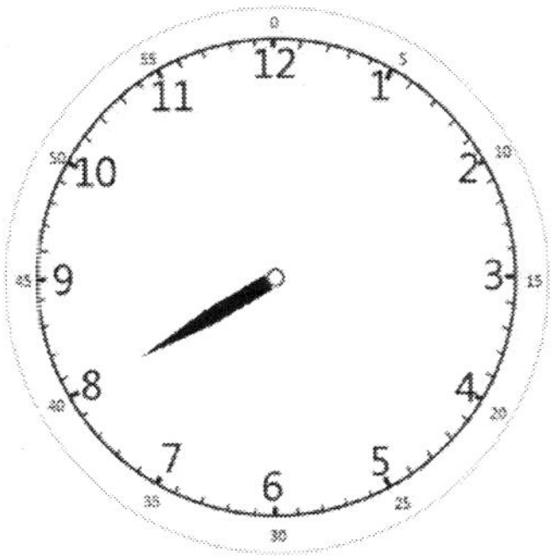  

________________ ________________ ________________

## Module 7 Practice Lesson 2

5. Connect the correct times with the clock faces.

4:00 2:00 9:00 7:00

Seven o'clock Nine o'clock Two o'clock Four o'clock

6. Connect the correct times with the clock faces.

12:30 4:30 10:30 2:30

Half-Past Two Half-Past Four Half-Past Twelve Half-Past Ten

7. Write down the correct times for the following clock faces. The first question has been done for you.

| | | |
|---|---|---|
|  |  |  |
| 1:00<br>One o'clock | | |

| | | |
|---|---|---|
|  |  |  |
| | | |

8. Draw the hour and minute hands to show the correct time.

| | | |
|---|---|---|
| 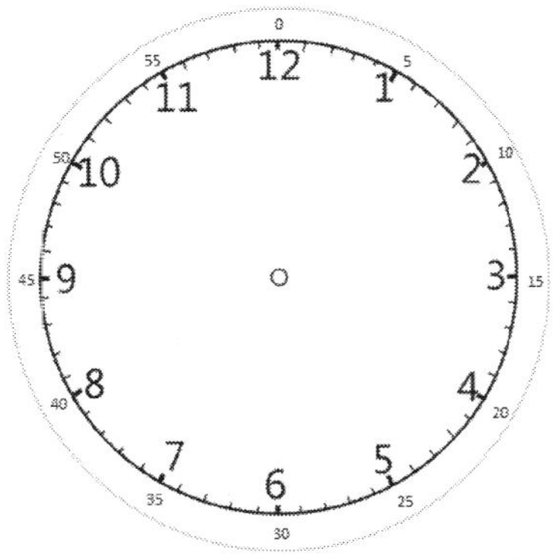 | 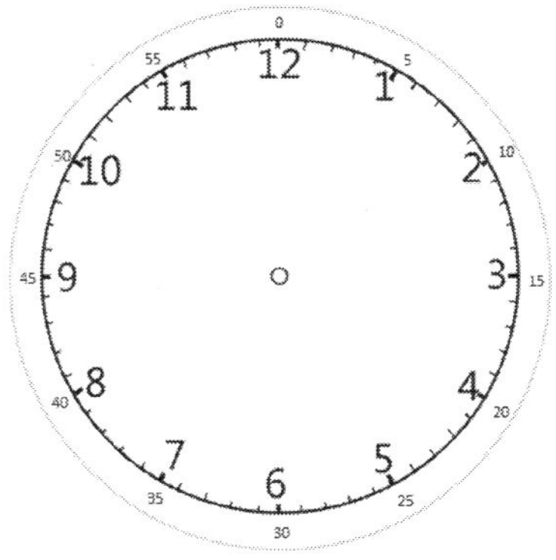 | 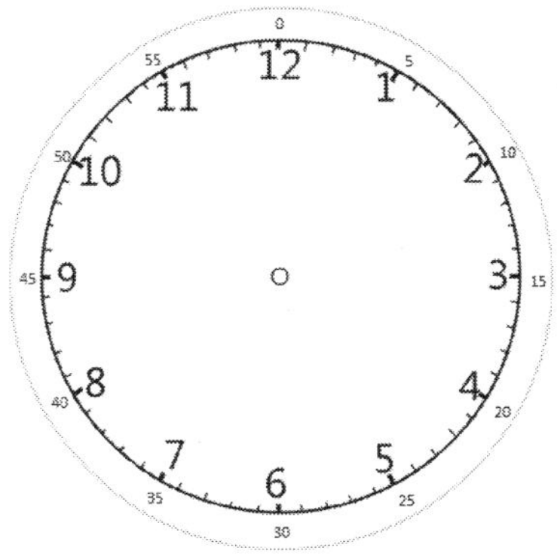 |
| 5:00 | 8:00 | 10:00 |

## Module 7 Practice Lesson 3

9. Connect the correct time to the clock faces.

| | | | |
|---|---|---|---|
| 7:30 | 9:30 | 5:30 | 3:30 |

| | | | |
|---|---|---|---|
| Half-Past Five | Half-Past Three | Half-Past Nine | Half-Past Seven |

10. Connect the correct time to the clock faces.

| | | | |
|---|---|---|---|
| 6:30 | 1:30 | 8:30 | 11:30 |

| | | | |
|---|---|---|---|
| Half-Past One | Half-Past Six | Half-Past Eleven | Half-Past Eight |

11. Write down the correct time for the following clocks. The first question has been done for you.

| | | |
|---|---|---|
|  |  |  |
| 10:30<br>Half-Past Ten | | |

| | | |
|---|---|---|
|  |  |  |
| | | |

12. Draw the hour and minute hands to show the correct time.

| | | |
|---|---|---|
| 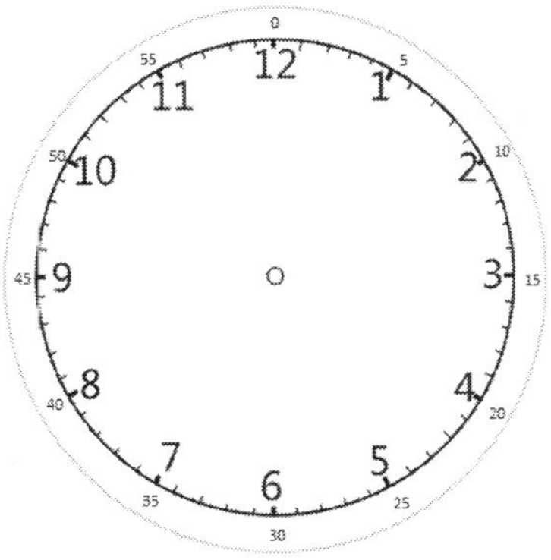 | 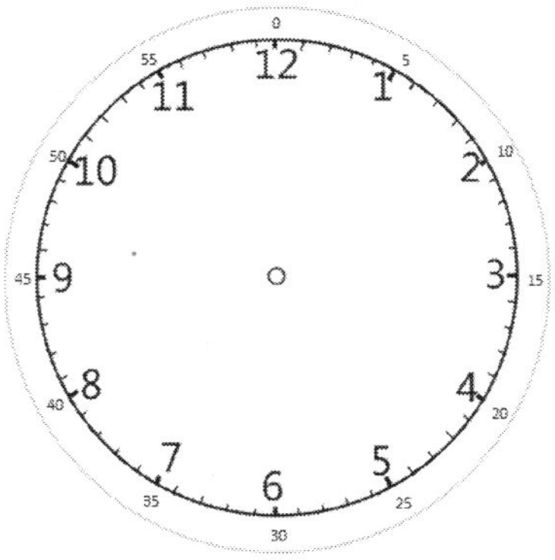 | 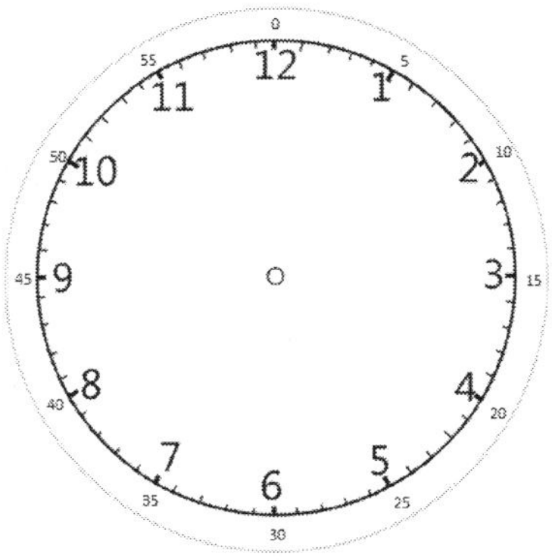 |
| 1:30 | 8:30 | 11:30 |

## Module 7 Practice Lesson 4

7. Complete the daily schedule for Harry with the given time.

| *Time* | *Things to do* |
|---|---|
| | Get up |
| | Have breakfast |
| | Go to work |
| | Break |
| | Have lunch |
| | Back to work |
| | Back home |
| | Have dinner |
| | Exercise |
| | Go to bed |

| *Choose time below* |
|---|
| 12:00 in the afternoon |
| 5:00 in the afternoon |
| 7:00 in the afternoon |
| 3:00 in the afternoon |
| 7:30 in the morning |
| 8:00 in the morning |
| 9:00 in the afternoon |
| 12:30 in the afternoon |
| 8:30 in the morning |
| 11:30 in the morning |

8. George is drawing hour and minute hands on the clock. But he missed some hands for each clock. Can you try to fix them?

| | | |
|---|---|---|
|  |  |  |
| 3:00 | 6:00 | 11:00 |

| | | |
|---|---|---|
|  | 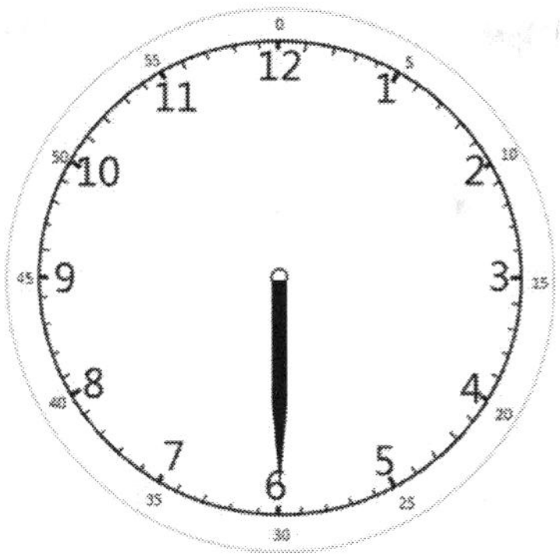 |  |
| 8:30 | 11:30 | 3:30 |

9. What is the new time of each clock after 2 hours? The first question has been done for you.

| | | |
|---|---|---|
|  |  |  |
| After 2 hours, it will be <u>11:00</u> | After 2 hours, it will be ______ | After 2 hours, it will be ______ |

10. Circle the correct answers to describe the following times. The first question has been done for you.

| | | | |
|---|---|---|---|
|  |  |  |  |
| It is almost 11:00<br>(It is alreay past 11:00) | It is almost 5:00<br>It is alreay past 5:00 | It is almost 9:00<br>It is alreay past 9:00 | It is almost 7:00<br>It is alreay past 7:00 |

## Module 8 20 以内进位加法 Addition with carryover (0 – 20)

In the previous chapters, we learned simple number addition. Such as, 5 + 3, 6 + 2, etc. Today we are going to learn some difficult addition problems, which will play a great role in our future learning.

❖ **进位的认识** **Knowing carryover**

Carryover can be seen as regrouping numbers into 10. For example, we can regroup 9 + 2 as 9 + 1 + 1. We know 9 + 1 = 10 and we need to plus 1 again, which is 11.

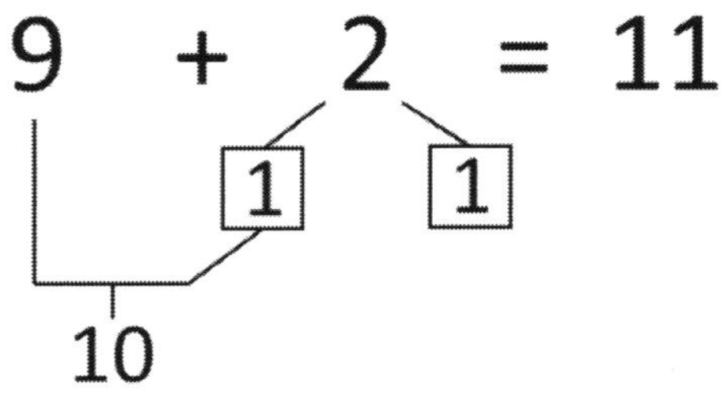


Another example: we can regroup 7 + 5 as 7 + 3 + 2. We know 7 + 3 = 10 and we need to plus 2 on 10 to get 12.

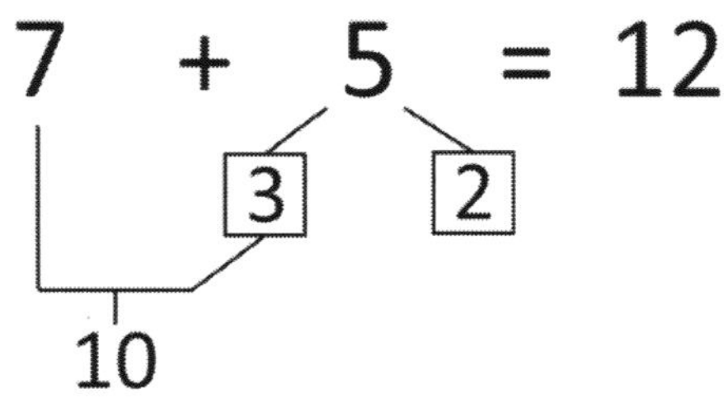

- 加法两式 **Addition laws (commutative law)**

Imagine you have **8 pencils**; your teacher gives you another **3 pencils**. How many pencils do you have?

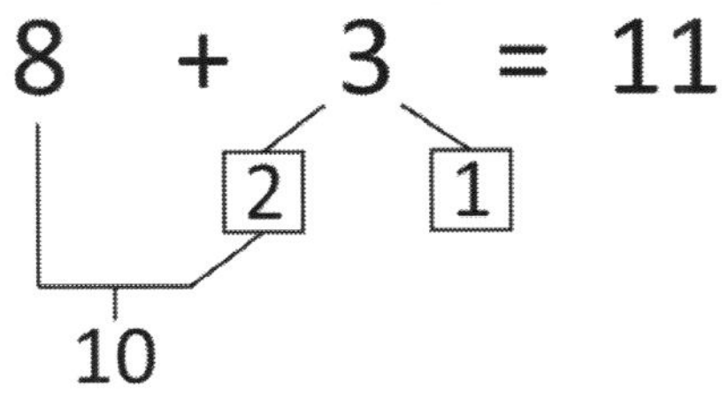


Now, imagine you have **3 pencils**; your teacher gives you another **8 pencils**. How many pencils do you have?

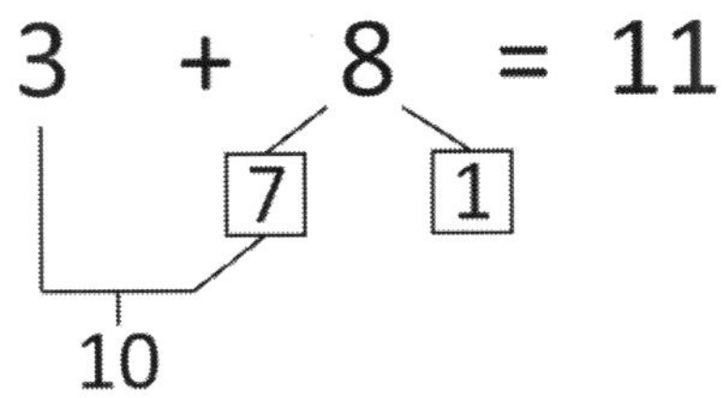


What have you found in these two questions?

□ ○ □ = □　　　□ ○ □ = □

In addition number sentence, two numbers can swap places and the answer (sum) won't change.

## Module 8 Practice Lesson 1

12. Write the correct number into the box.

a)

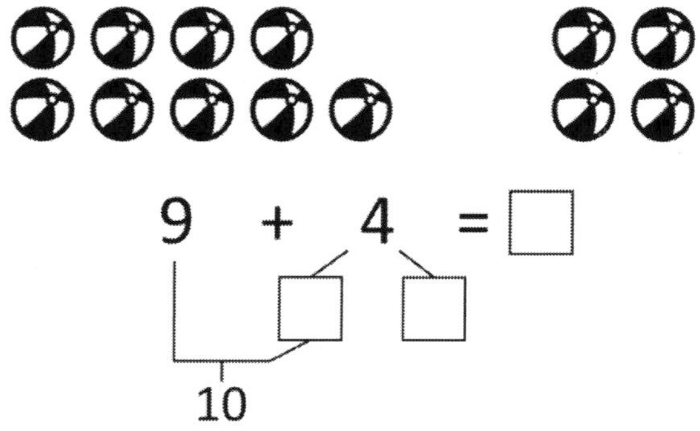


b)

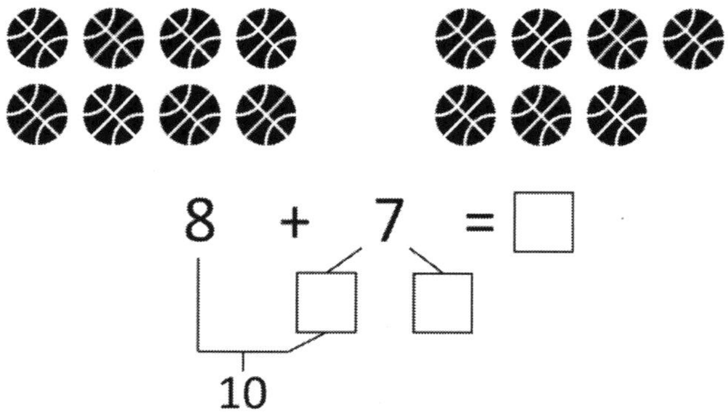


c)

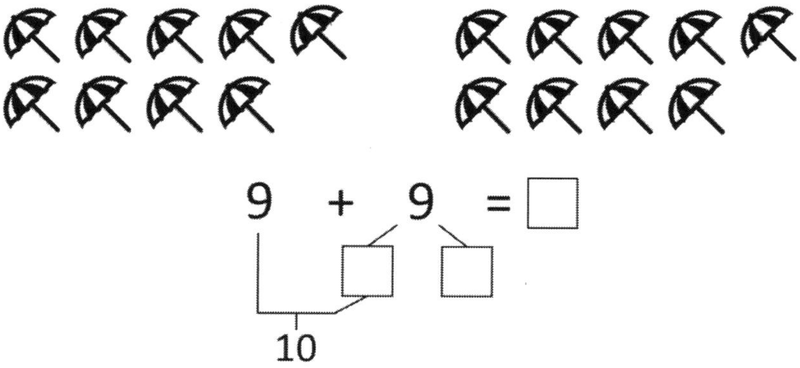


13. Find the answer for each number sentence and see what you have found.

| | |
|---|---|
| 9 + 1 + 4 =<br>9 + 5 = | 8 + 2 + 5 =<br>8 + 7 = |
| 9 + 1 + 6 =<br>9 + 7 = | 3 + 7 + 2 =<br>3 + 9 = |
| 9 + 1 + 8 =<br>9 + 9 = | 7 + 3 + 3 =<br>7 + 6 = |

14. Connect the chicken into correct house.

| | | |
|---|---|---|
| 7 + 3 + 4 | 13 | 8 + 7 |
| 4 + 6 + 3 | 14 | 7 + 7 |
| 8 + 2 + 7 | 15 | 8 + 9 |
| 8 + 2 + 5 | 17 | 4 + 9 |

15. Move number 9 along the chart to show the correct sum.

9 

+

| 1 | 2 | 3 | 4 | 5 | 6 | 7 | 8 | 9 | 10 |
|---|---|---|---|---|---|---|---|---|---|
| | | | | | | | | | |

16. There are 7 boats in the lake. Another 6 boats are joining in the group. How many boats are in the lake now?

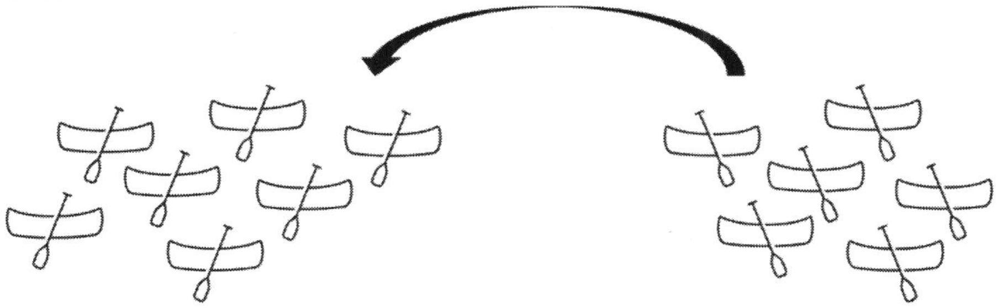

□ ○ □ = □ □ ○ □ = □

9. Calculate the following number sentences.

| | | |
|---|---|---|
| 9 + 3 = | 7 + 5 = | 8 + 4 = |
| 3 + 8 = | 6 + 7 = | 4 + 9 = |

10. Compare the numbers. Using “>”, “<” or “=”.

| | | |
|---|---|---|
| 9 + 7 ______ 15 | 8 + 5 ______ 10 | 7 + 6 ______ 15 |
| 3 + 8 ______ 13 | 7 + 7 ______ 14 | 3 + 9 ______ 10 |
| 2 + 9 ______ 12 | 5 + 8 ______ 18 | 7 + 4 ______ 11 |
| 8 + 9 ______ 14 | 7 + 5 ______ 13 | 6 + 9 ______ 15 |

11. Complete the table below.

8 →

\+

| | | | | | | | | | |
|---|---|---|---|---|---|---|---|---|---|
| 1 | 3 | 7 | 6 | 9 | 8 | 2 | 5 | 4 | 10 |
| | | | | | | | | | |

7 →

\+

| | | | | | | | | | |
|---|---|---|---|---|---|---|---|---|---|
| 5 | 7 | 10 | 3 | 2 | 6 | 8 | 4 | 1 | 9 |
| | | | | | | | | | |

12. Lead those students back home.

|  |  |  |
|---|---|---|
| 6 + 5 | 11 | 8 + 5 |
| 8 + 7 | 12 | 7 + 5 |
| 4 + 8 | 13 | 6 + 9 |
| 7 + 6 | 15 | 7 + 4 |

13. Draw the correct number of shapes for each question below.

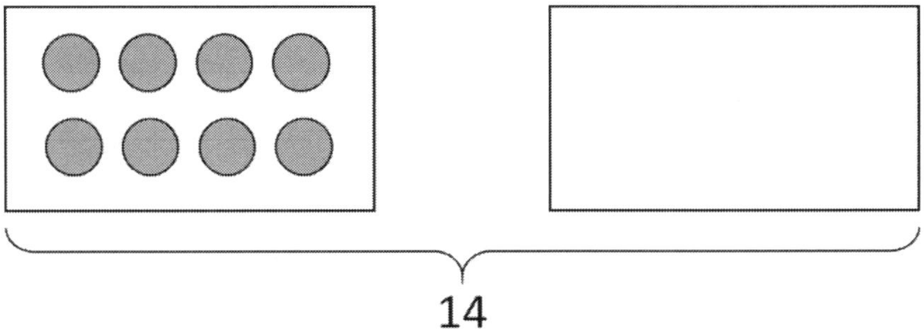


8 + $\square$ = 14

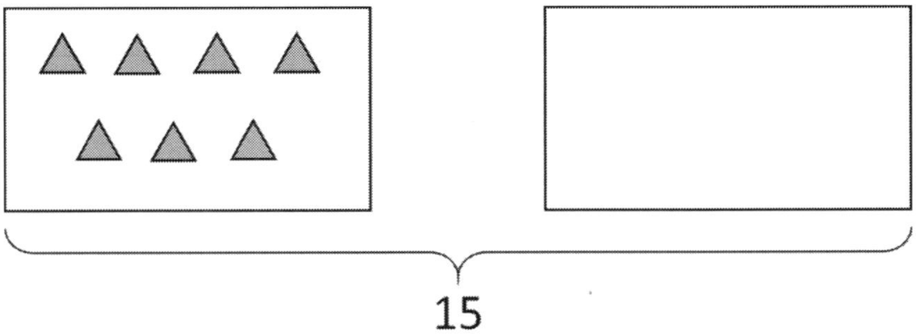


7 + $\square$ = 15

## Module 8 Practice Lesson 3

**13.** Fill in the correct number to make the number sentence true.

| | | |
|---|---|---|
| $6 + \square = 14$ | $8 + \square = 16$ | $7 + \square = 15$ |
| $5 + \square = 12$ | $9 + \square = 17$ | $4 + \square = 12$ |
| $3 + \square = 11$ | $5 + \square = 14$ | $4 + \square = 13$ |

14. A tower of bricks is built so that each number of two bricks add up to have a number of new bricks placed on top. Fill in the blank space.

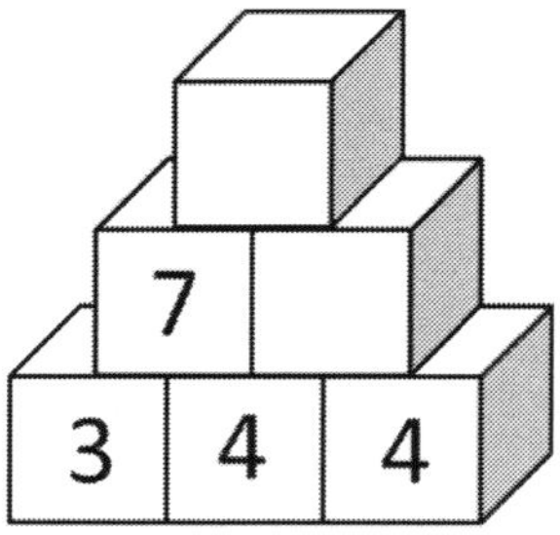

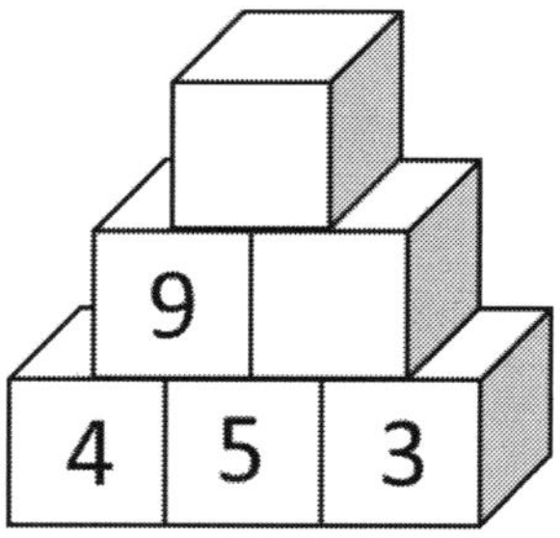

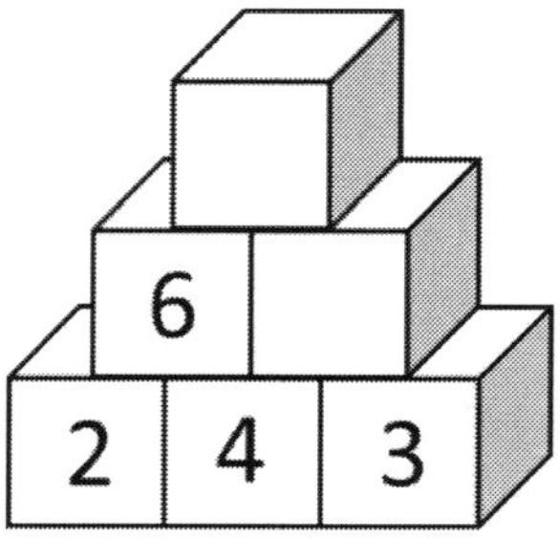


15. Make up your own number sentence, make sure the sum of each number sentence must be the same as the given one.

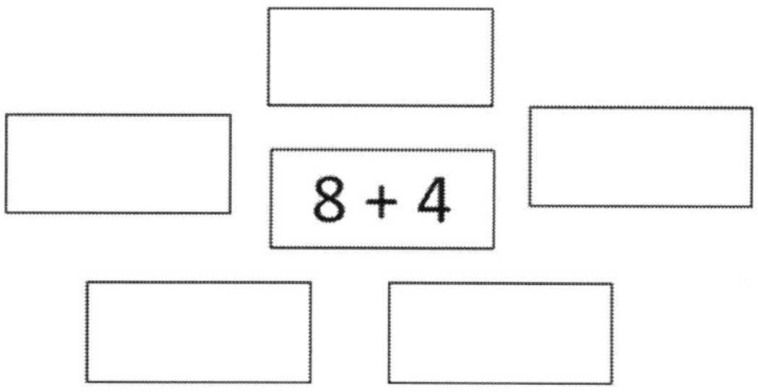

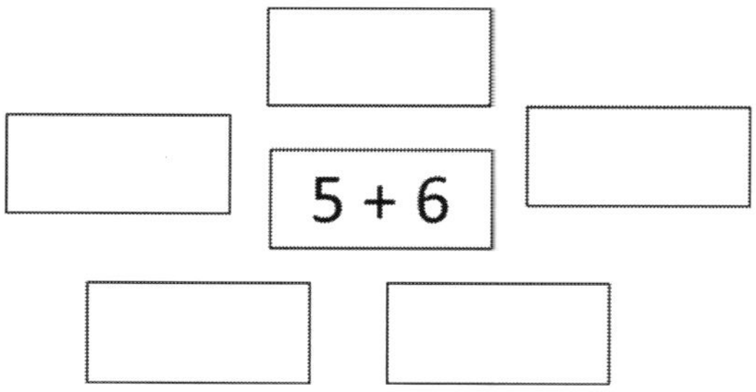


16. Put 2, 3, 4, 5, 6, 7, 8, 9 in these boxes to make the number sentence true.

$\square + \square = \square + \square = \square + \square = \square + \square$

17. Jonathan has $8 in his pocket. His mother gives him another $9 because he got 100 marks on his maths test. How much money does he have altogether?

18. Harry has 7 marbles. He gives 2 marbles to his friend and his brother gives 9 marbles to Harry. How many marbles does Harry have altogether?

19. There are 14 students in the classroom. Each student can get one apple. Which two plates need to be chosen? Please circle the correct plates.

20. How many fish did the cat caught?

## Module 8 Practice Lesson 4

11. Write addition (+) or subtraction (–) sign to make the following number sentence true.

6 ◯ 6 = 12     6 ◯ 7 = 13     9 ◯ 6 = 3

7 ◯ 7 = 0     9 ◯ 3 = 12     12 ◯ 2 = 10

12. Find the pattern of the following question and complete in the missing number.

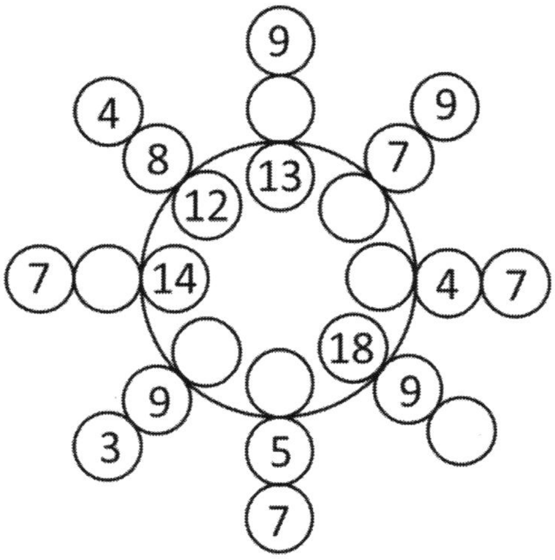


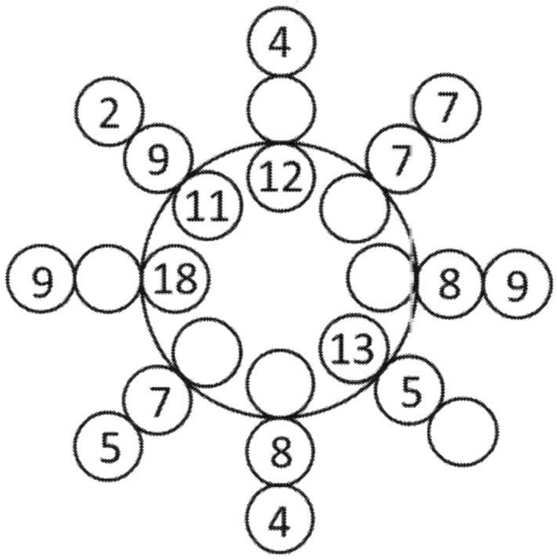


13. Write the correct number sentence and solve them.

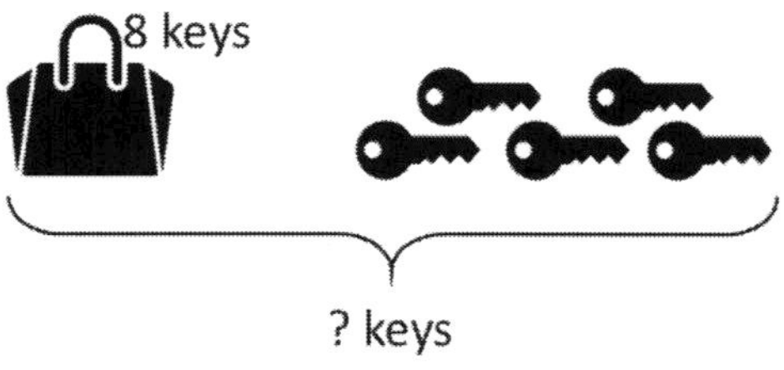


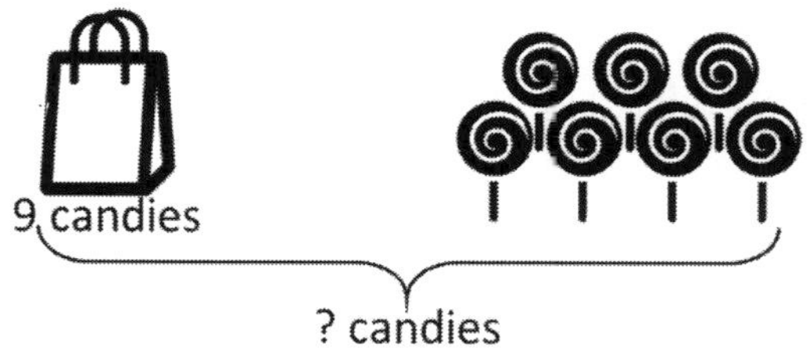


☐ + ☐ = ☐     ☐ + ☐ = ☐

☐ + ☐ = ☐     ☐ + ☐ = ☐

14. There were a total of 8 passengers on the bus. 9 people get on the bus at the first stop. 6 passengers get off the bus at the second stop. How many passengers are still on the bus?

15. Serena has 8 fish in the water tank. Her father bought another 7 fish for her. How many fish are in the tank now?

Who's working out is correct? Give a reason.

16. How many students are in the line?

## Module 9 分类与整理 Data collection and sorting

### ❖ 找不同 Odd ones out

The odd one out is the person or thing that is different from the other members of a group.

For example:

In the above student group, only Jessy has the letter start with J, therefore Jessy is the odd one out.

### ❖ 分组 Sorting

Sorting is rearranging items in some similar properties. These can be either the same colour, same shape or same value.

For example:

We can sort the animals above into two groups: animals that have 4 legs and animals that have less than 4 legs:

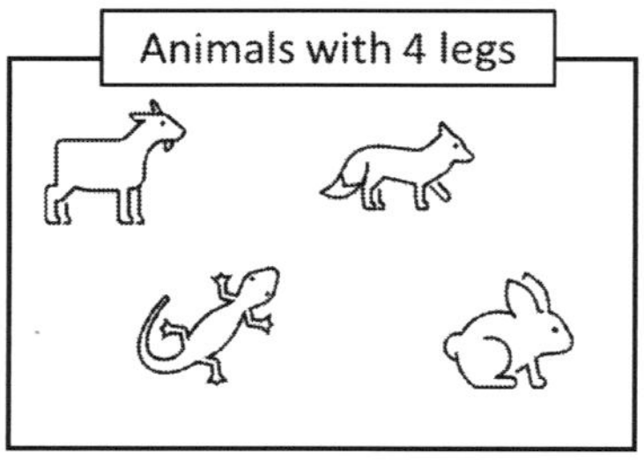


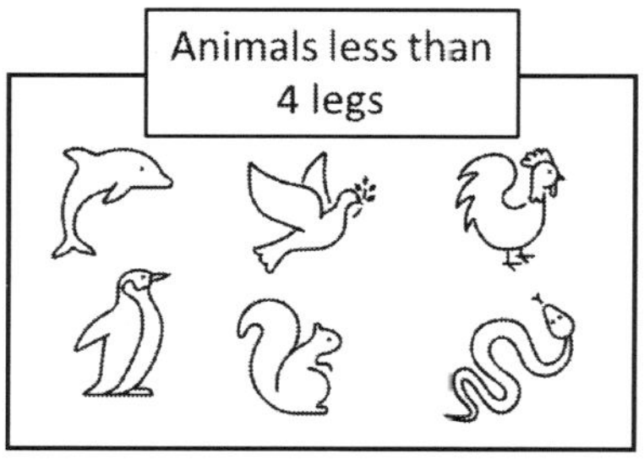


There is more than one way to sort the animals above. We can also sort them by size or species as well.

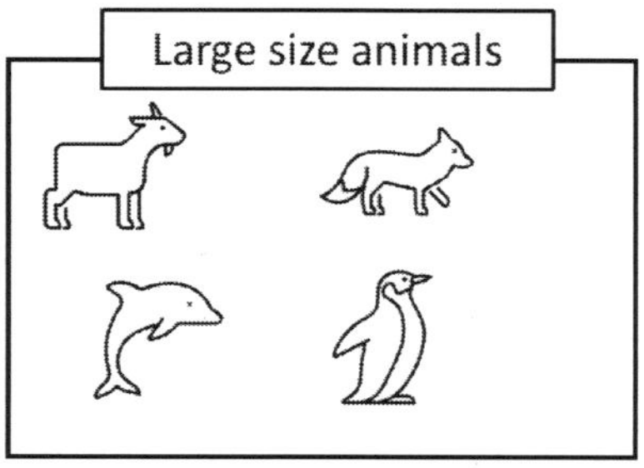


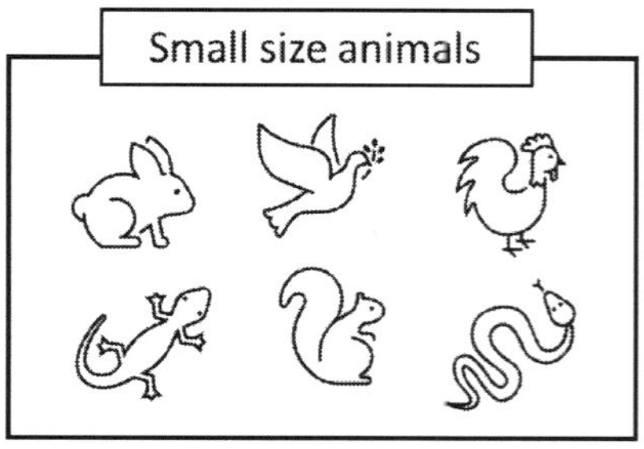

## 数据的记录 Recording data

Data is used to describe a group of numbers, objects or measurements.
Data can be collected in many ways, such as observations, surveys and interviews.
The purpose of data recording is to get information about a specific topic, which can help in making decisions.

Example: Count the following shapes and record data in the table.

| Shape |  |  |  |
|---|---|---|---|
| The number of shapes | 6 | 5 | 4 |

### Tally marks

Sometimes we might easily forget which number we are up to. Therefore we can use **tally marks** to help us keep track of numbers. For the first four numbers, we will draw a vertical line for each number. For the fifth number, we draw a diagonal line across the previous four numbers, this line represents the fifth number.

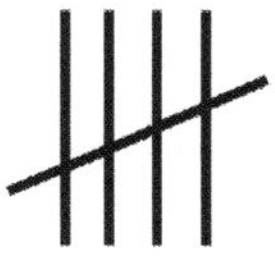

Example: Count the following shapes and record data in the table by using tally marks.

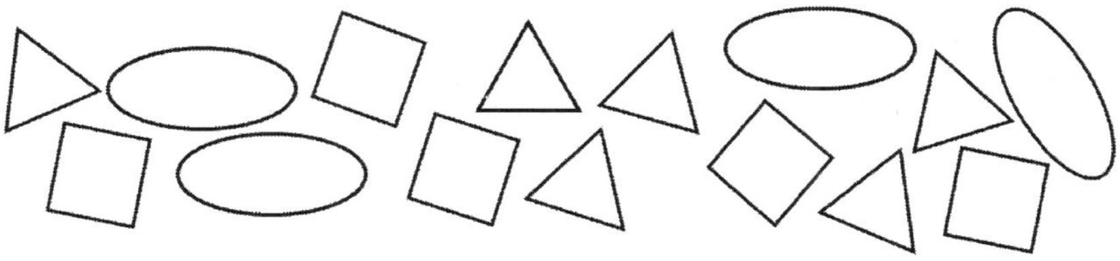

| Shape |  |  |  |
|---|---|---|---|
| The number of shapes | 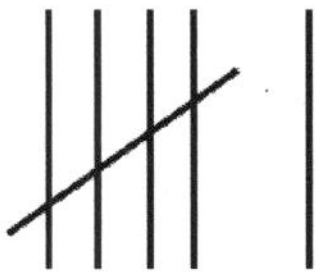 | 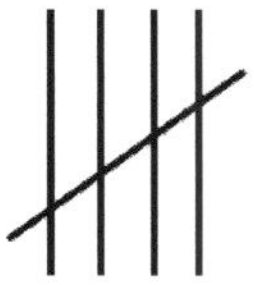 |  |

## Module 9 Practice Lesson 1

17. Colour the animals in the following diagram.

There are __________ animals in the diagram.

18. Circle the one of the pictures is different from the other three.

a)

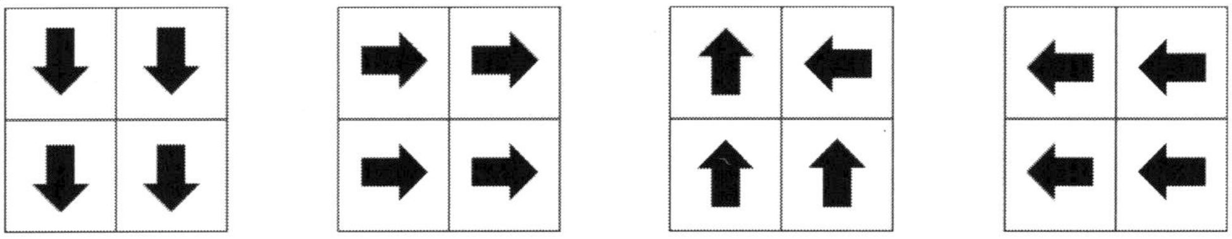

b)

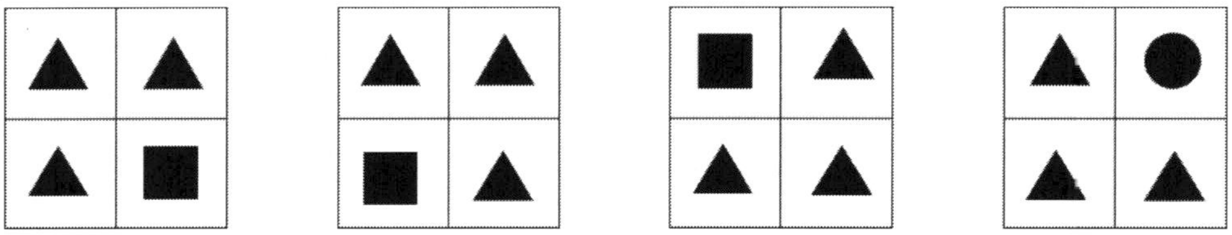

c)

| | |
|---|---|
| 1 | 2 |
| 3 | 4 |

| | |
|---|---|
| 1 | 3 |
| 2 | 4 |

| | |
|---|---|
| 1 | 4 |
| 2 | 3 |

| | |
|---|---|
| 3 | 2 |
| 4 | 1 |

d)

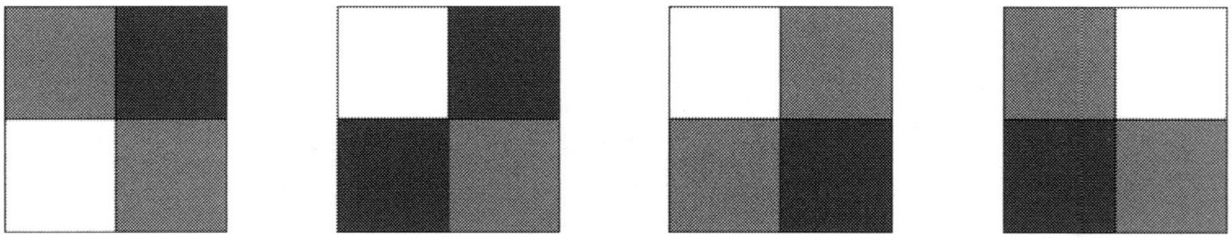

19. Sort the following items into two groups.

20. Count how many of each flower and complete the table below.

| Flower |  |  |  |
|---|---|---|---|
| Number of flowers | | | |

a) How many flowers are there in total?

b) Which flower has the most amount?

## Module 9 Practice Lesson 2

14. Circle the one of the pictures is different from the other three. And tell your teacher the reason.

15. Sorting the following icons into two groups.

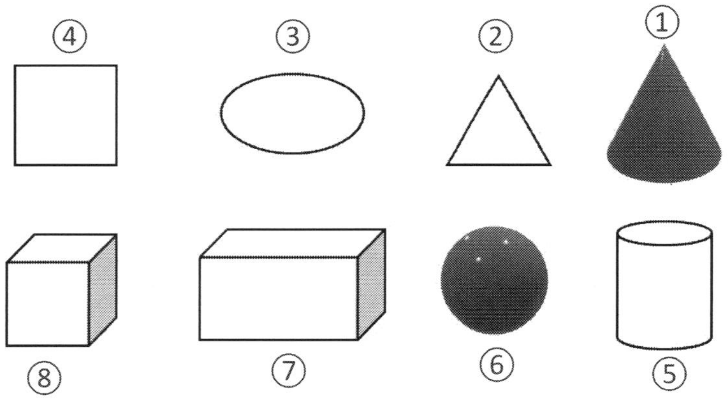


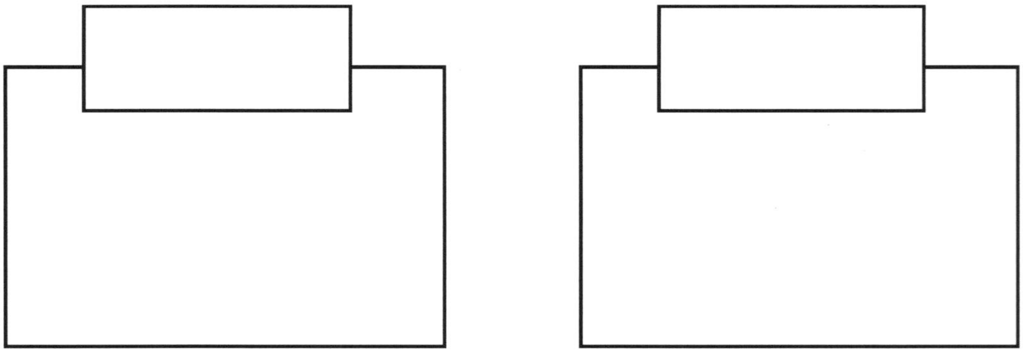

16. There are 7 differences in the two photos. Circle the differences.

17. Count the following shapes and complete the table below.

| Shapes |  |  |  |
|---|---|---|---|
| Number of shapes | | | |

a) How many shapes are there in total?

b) Which shape has the most amount?

c) Which shape has the least amount?

d) What is the difference in amount between the Sun and the Star?

## Module 9 Practice Lesson 3

**21.** In each row, circle one of the pictures that is different from the other three. Tell your teacher the reason why you chose that picture.

**22.** Ming wants to know the favourite colours of the students in his class. Therefore, he asked his students, and the result is shown below.

Red, Yellow, Blue, Yellow, Green, Blue, Red, Red, Red, Yellow, Yellow, Blue, Red, Yellow, Blue, Yellow, Yellow, Green, Blue, Blue, Blue, Green, Green, Blue, Blue, Blue, Green, Blue

| Colour | Red | Blue | Green | Yellow |
|---|---|---|---|---|
| Number of students | | | | |

a) How many students are there in the Ming's class?

b) Ming wants to design a uniform for his class, which colour should he consider first?

**23.** The following diagram shows the favourite food of each student. Use tally marks to complete the table.

| **Food** |  |  |  |  |
|---|---|---|---|---|
| **Tally mark** | | | | |
| **Number of students** | | | | |

a) How many students are in the class?

b) Which food is the most preferred?

c) Which two foods are favoured by the same number of students?

**24.** In the previous chapter, we have learnt about squares and rectangles. List one similarity and difference between these two shapes.

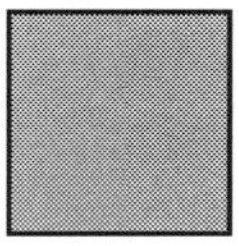

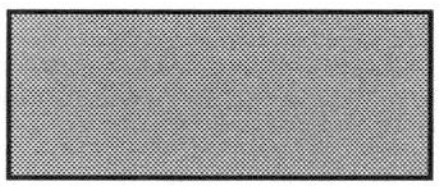

| Similarity | Difference |
|---|---|
| | |

## Module 9 Practice Lesson 4

17. Circle the one of the pictures is different from the rest. And tell your teacher the reason.

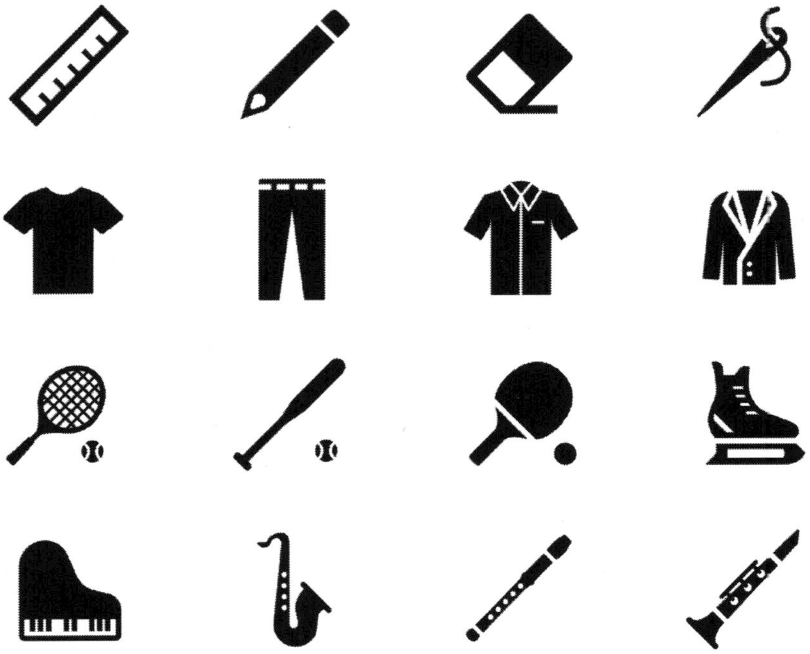

18. Sort the following clothing and complete tables below.

a) If we want to group them by names.

| | Shirt | Pants | Skirt |
|---|---|---|---|
| Amount | | | |

b) If we want to sort them only in two groups. How can we sort them?

| | | |
|---|---|---|
| Amount | | |

19. Mr. Fred want to know students' favourite sports. Here is what he found out:

a) How many types of sports are there in total?

b) Use tally marks to complete the table below.

| Sport | Cricket | Skiing | Karate | Swimming | Tennis | Soccer |
|---|---|---|---|---|---|---|
| Tally marks | | | | | | |

c) Use numbers to complete the table below.

| Sport | Cricket | Skiing | Karate | Swimming | Tennis | Soccer |
|---|---|---|---|---|---|---|
| Number of students | | | | | | |

d) Which sport is the most favoured?

e) Ask your classmates what their favourite sport is and tell your teacher.

## Module 10 20 以内的退位减法 Subtraction with borrowing (0 – 20)

In the previous chapters, we learned simple number subtraction. Such as 9 – 3, 5 – 1, etc. Today we are going to learn some difficult subtraction problems, which will play a great role in our future study.

### ❖ 退位的认识 Knowing borrowing

Borrowing can also be called as "trading". Borrowing applies when you are subtracting one number that is greater than another.

For example: 19 – 5 would not need borrowing, as 9 minus 5 we know is 4.
12 – 5 would use borrowing, as 2 cannot minus 5.

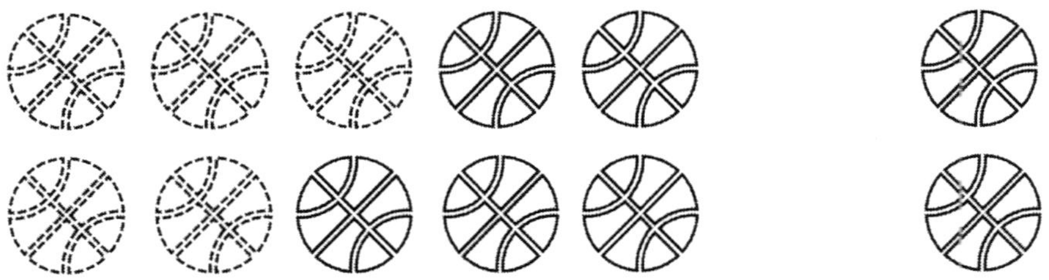

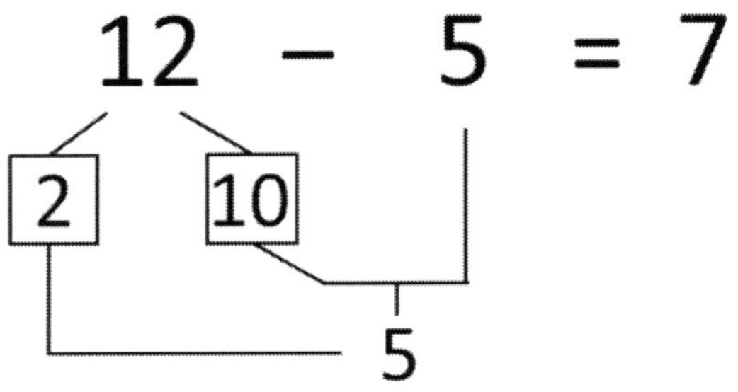


Another example: we can subtraction 14 – 9. Split 14 as 10 and 4, subtract 9 from 10 will get 1. And use 1 to plus 4 to get the answer.

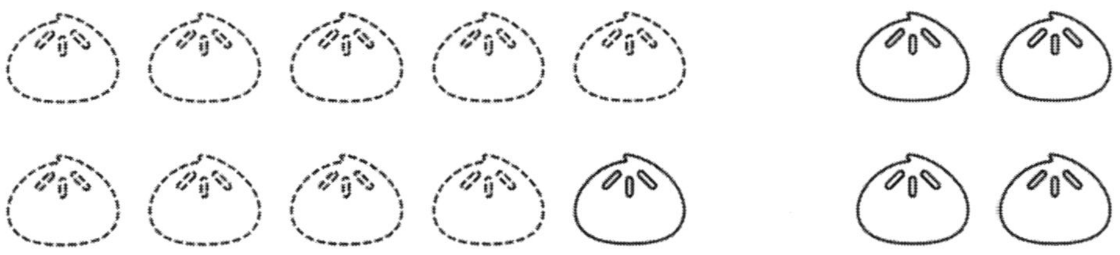

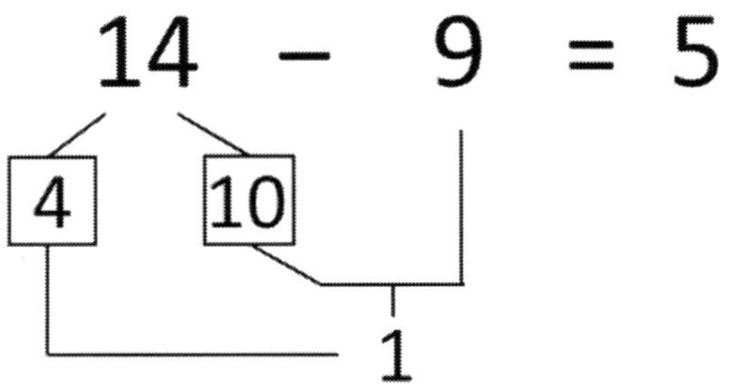

- **加减法的关系 Relationship between addition and subtraction**

In module 8, we have learned commutative law, which 8 + 3 = 11 and 3 + 8 = 11. Now what does 11 – 3 and 11 – 8 equals to?

$$11 - 3 = 8$$

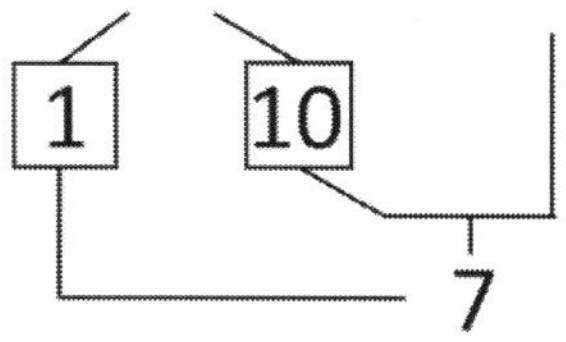


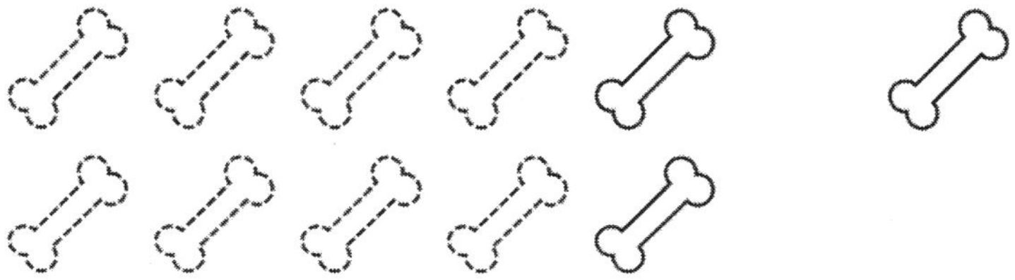

$$11 - 8 = 3$$

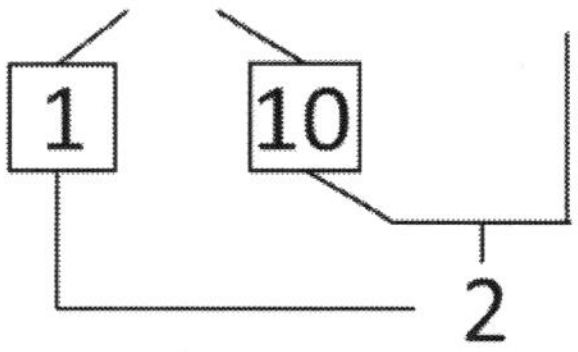


Now, what have you found?

Can you complete the rest two number sentences by using number '7', '8' and '15'?

| | | |
|---|---|---|
| $8 + 7 = 15$ | | □ – □ = □ |
| $7 + 8 = 15$ | | □ – □ = □ |

## Module 10 Practice Lesson 1

21. Circle the amount that you need to take away and find the final answer.

| | |
|---|---|
|  15 – 8 = |  12 – 5 = |
| 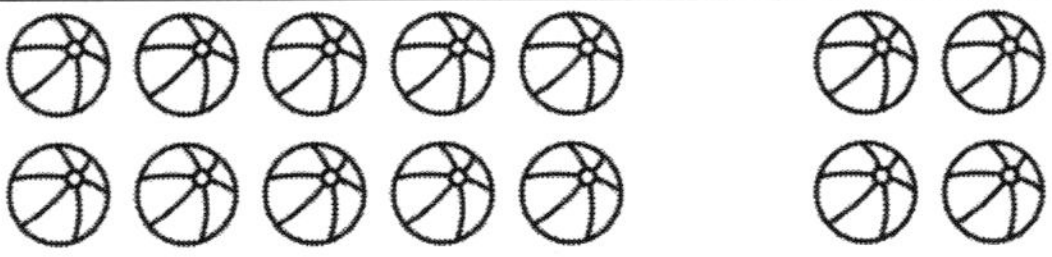 14 – 7 = | 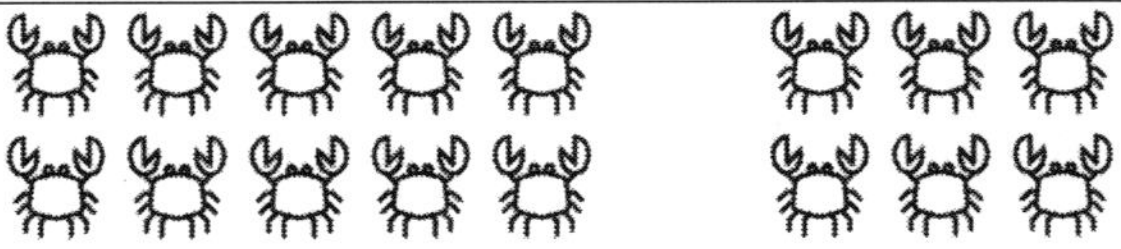 16 – 9 = |

22. Calculate.

| | | | |
|---|---|---|---|
| 11 – 9 = | 12 – 9 = | 14 – 9 = | 17 – 9 = |
| 12 – 8 = | 13 – 8 = | 15 – 8 = | 17 – 8 = |
| 10 – 7 = | 12 – 7 = | 14 – 7 = | 16 – 7 = |

23. Connect the dogs with correct bones.

| | | | | | |
|---|---|---|---|---|---|
|  |  |  |  |  |  |
| 15 – 9 | 11 – 4 | 17 – 8 | 14 – 5 | 14 – 7 | 12 – 6 |

| | | |
|---|---|---|
|  |  |  |
| 6 | 7 | 9 |

24. Move the number 9 towards each number to find their difference.

| | | | | | | | | | |
|---|---|---|---|---|---|---|---|---|---|
| 13 | 11 | 12 | 16 | 14 | 15 | 17 | 18 | 19 | 10 |

−

| |
|---|
| 9 |

→

| | | | | | | | | | |
|---|---|---|---|---|---|---|---|---|---|
| | | | | | | | | | |

25. Find the relationship between the addition and subtraction of the following diagrams.

a)

8 + 3 = 11 □ − □ = □

□ + □ = □ □ − □ = □

b)

7 + 6 = 13 □ − □ = □

□ + □ = □ □ − □ = □

c)

6 + 9 = 15 □ − □ = □

□ + □ = □ □ − □ = □

## Module 10 Practice Lesson 2

18. Connect the books with correct backpack.

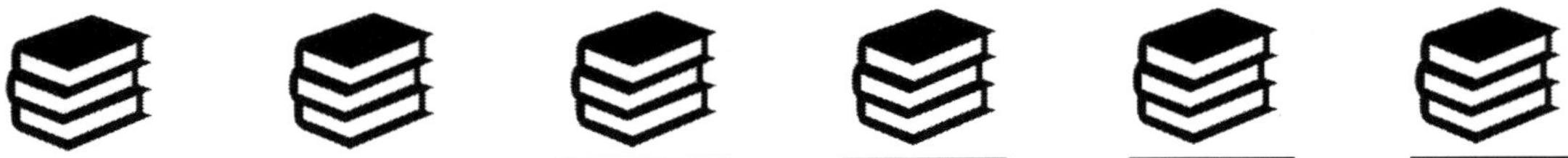

| | | | | | |
|---|---|---|---|---|---|
| 15 – 8 | 12 – 7 | 13 – 7 | 17 – 9 | 11 – 7 | 14 – 5 |

| | | | | | |
|---|---|---|---|---|---|
| 9 | 8 | 4 | 5 | 6 | 7 |

19. Fill in the missing number in each number sentence and tell your teacher what you have found.

| | |
|---|---|
| 6 + ☐ = 15<br>15 – 9 = ☐ | 7 + ☐ = 12<br>12 – 7 = ☐ |
| 9 + ☐ = 14<br>14 – 9 = ☐ | 8 + ☐ = 16<br>16 – 8 = ☐ |

20. Compare the numbers with “>”, “<” and “=”.

| | | |
|---|---|---|
| 18 – 7 ______ 5 | 17 – 9 ______ 7 | 11 – 5 ______ 6 |
| 13 – 8 ______ 5 | 12 – 7 ______ 3 | 14 – 7 ______ 8 |
| 18 – 9 ______ 7 | 13 – 7 ______ 7 | 16 – 7 ______ 8 |
| 11 – 5 ______ 4 | 11 – 8 ______ 3 | 12 – 5 ______ 4 |

21. Complete the table below. The first question has been done for you.

| | 11 | 13 | 14 | 12 | 15 | 16 | 17 |
|---|---|---|---|---|---|---|---|
| – 8 | 3 | | | | | | |
| – 9 | 2 | | | | | | |

22. A tower of bricks is built so that each number of two bricks add up to have a number of new bricks placed on top. Fill in the blank space.

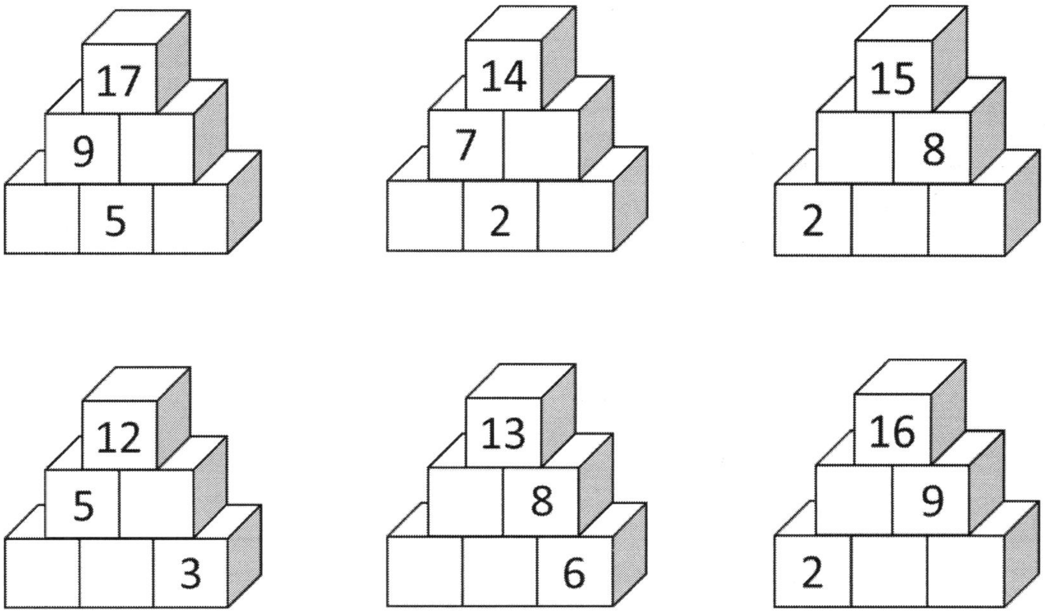


23. Put number '1', '2', '4', '5' and '7' in the missing boxes to make the sum of each row, column and diagonal equals to 15.

| | | |
|---|---|---|
| | 9 | |
| | | 3 |
| 6 | | 8 |

**25.** Complete the missing number by solving each number sentence.

| | | | | | | |
|---|---|---|---|---|---|---|
| 12 | | | | 14 | | |
| 14 | − 7 = | | | 12 | − 9 = | |
| 16 | | | | 13 | | |
| 13 | | | | 11 | | |

**26.** Can you solve the following question by using the method below?

$11 - 8 = 11 - 1 - 7 = 3$

(8 split into 1 and 7)

$14 - 9 = \square - \square - \square = \square$

(9 split into □ and □)

$13 - 7 = \square - \square - \square = \square$

(7 split into □ and □)

**27.** Use your smart brain to find what is the value for each shape below.

6 + □ = 15 □ =

□ + △ = 12 △ =

△ + 8 = ○ ○ =

28. There were 14 carrots in the cage. If the rabbit ate 8 of them, how many carrots are left in the cage?

29. Jimmy and Anderson are playing basketball. Jimmy scores 12 points and Anderson scores 7 points. How many scores does Jimmy have more than Anderson?

30. There are 17 students in the classroom. 8 students are boy. How many girls are in the classroom?

20. Write the correct number sentence for each diagram below.

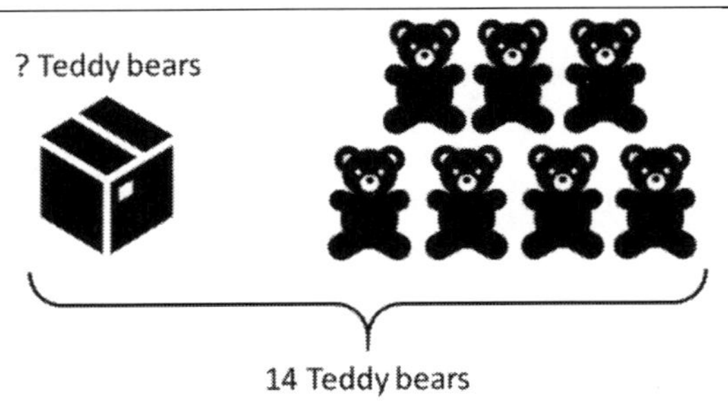


□ − □ = □

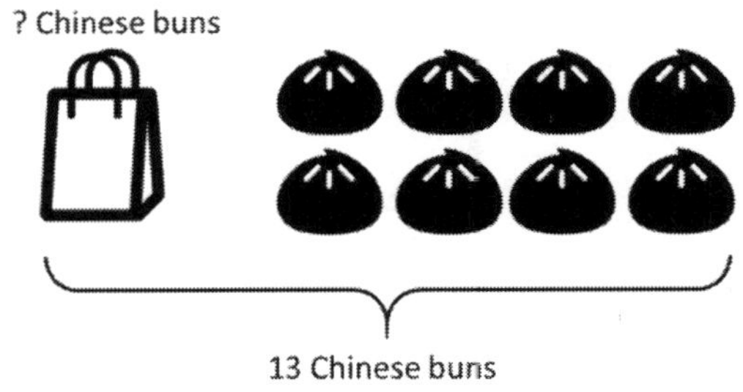


□ − □ = □

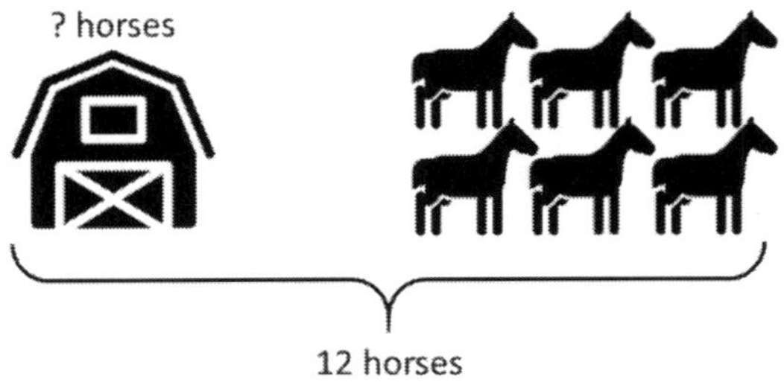


□ − □ = □

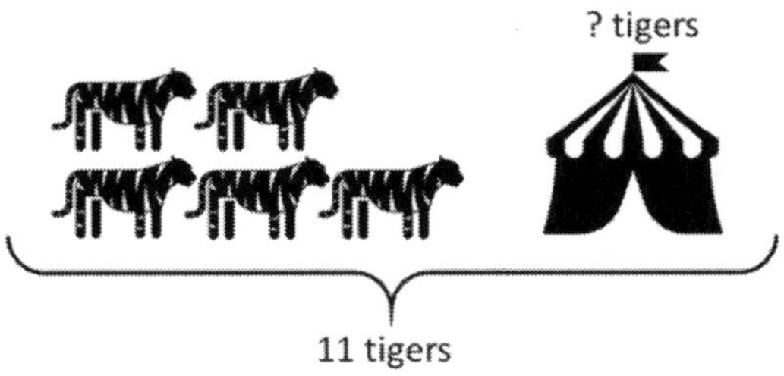


□ − □ = □

➢ **Question 2 – 5 refer to the following information**

*Do you know the Twelve Chinese Zodiac? There are 12 animals represents 12 different years. For example, year 2020 is the year of Rat, and year 2021 is the year of Ox.*

21. These 12 animals are having a party today, there are 7 animals are already at the party place. How many animals are still on their way to party place?

22. The seven animals already at the party are rate, ox, rabbit, dragon, snake, monkey and goat. What other animals are not present?

23. The Twelve Chinese Zodiac are in sequence. I have listed the order below.
If this year is the year of Ox, then next year will be the year of Tiger!
If this year is the year of dog, what animal after 4 years' time?

24. John is 12 years old, and he was born in the year of Snake. His little brother is 9 years younger than John. How old is his little brother? And what is his brother's zodiac?

## Module 11 数位的初步认识 Introduction to place value

### ❖ 数位表 Place value chart

Each number has place value. Place value means what is the number locates and how large is the number.

For example, 15 means we have 5 ones and 1 tens. Therefore, the 5 in the ones place and the 1 in the tens place.

We can express all the numbers on a place value chart. And there are many different types of place value charts around the world.

| Hundreds | Tens | Ones |
|---|---|---|
| 0 | 1 | 5 |

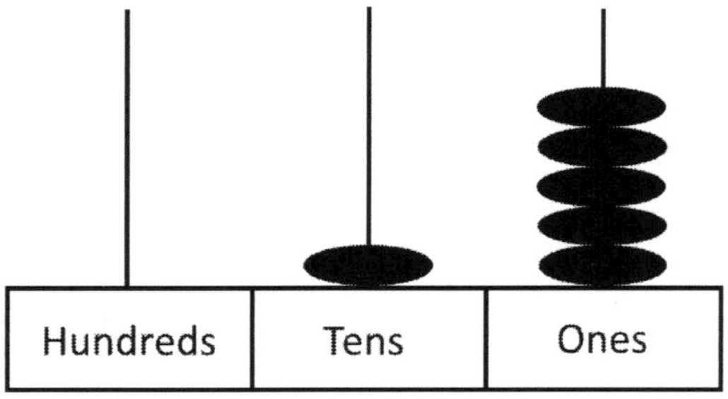


### ❖ 数位表示数字 Place value to show numbers

We can express any number on the place value chart if we understand the place value for each number.

For example, 142 in word is one hundred forty-seven. 1 in the hundreds place, 4 in the tens place and 2 in the ones place.

| Hundreds | Tens | Ones |
|---|---|---|
| 1 | 4 | 2 |

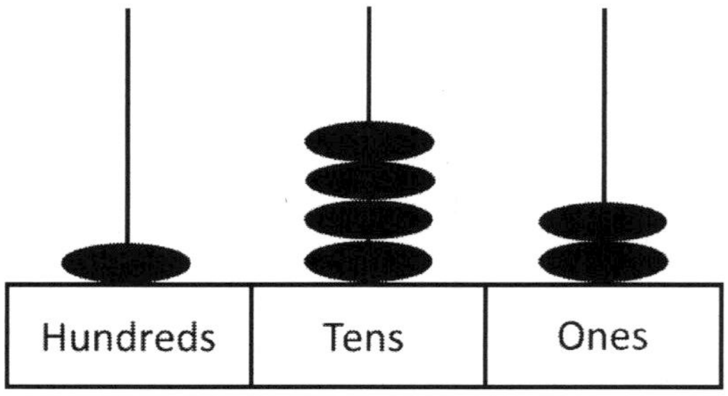


Another example, 109 in word is one hundred and nine. 1 in the hundreds place, 0 in the tense place and 4 in the ones place.

| Hundreds | Tens | Ones |
|---|---|---|
| 1 | 0 | 4 |

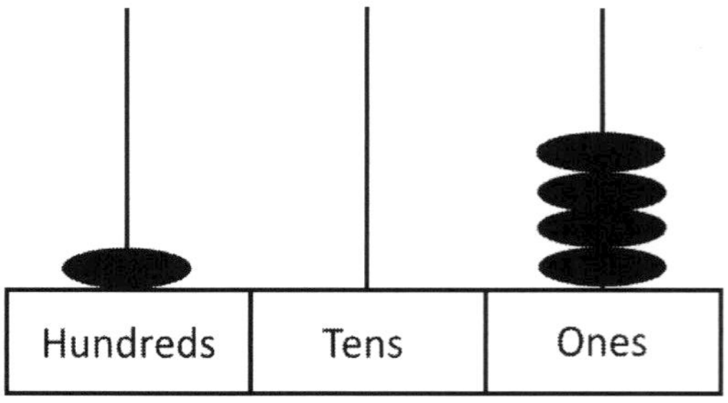

## Module 11 Practice Lesson 1

26. Count how many fruits are there.

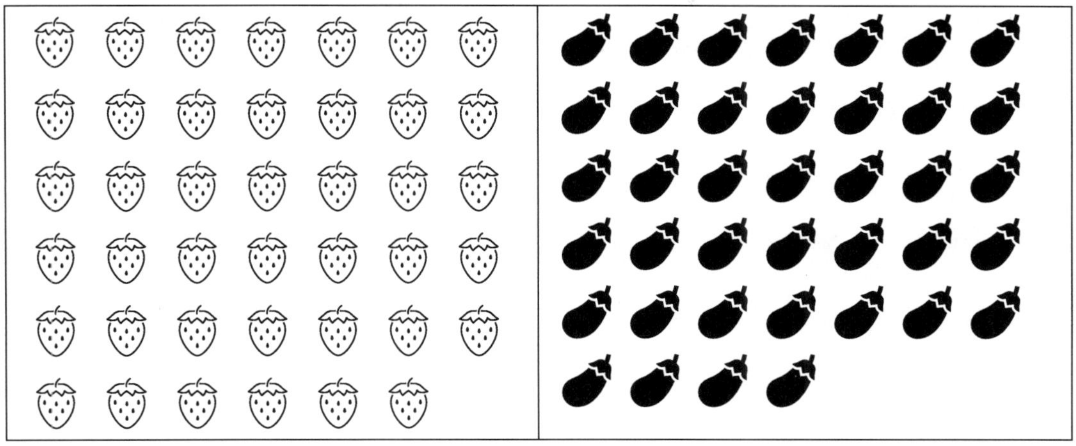

There are ____________ strawberries, and ______________ eggplants.

27. Write the next five numbers after the following numbers.

41, ___________, ___________, ___________, ___________, ___________

72, ___________, ___________, ___________, ___________, ___________

86, ___________, ___________, ___________, ___________, ___________

95, ___________, ___________, ___________, ___________, ___________

28. Write down the following numbers according to the place value chart.

| Hundreds | Tens | Ones |
|---|---|---|
| 0 | 3 | 4 |

____________

| Hundreds | Tens | Ones |
|---|---|---|
| 0 | 7 | 5 |

____________

| Hundreds | Tens | Ones |
|---|---|---|
| 0 | 8 | 0 |

____________

| Hundreds | Tens | Ones |
|---|---|---|
| 0 | 9 | 9 |

____________

| Hundreds | Tens | Ones |
|---|---|---|
| 1 | 0 | 2 |

____________

| Hundreds | Tens | Ones |
|---|---|---|
| 1 | 1 | 1 |

____________

29. Find the number without counting them one by one.

a)

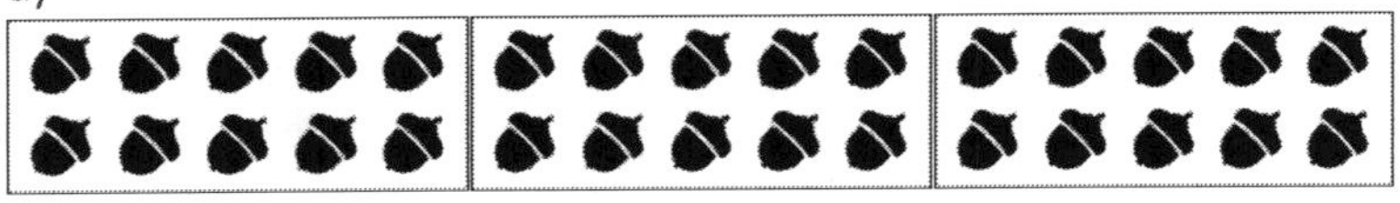

____________nuts

b)

____________books

c)

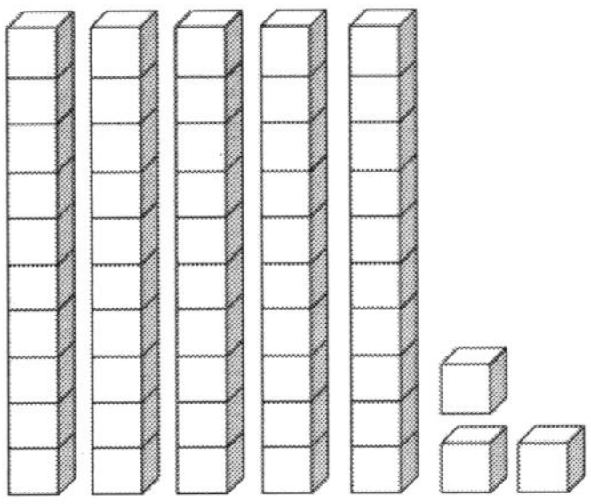

____________cubes

30. Find the numbers by reading the following statements.

a) I have 2 in the ones place and 1 in the tens place. What number am I?

b) I have 3 in the tens place and 0 in the ones place. What number am I?

c) I have 8 in the tens place and 3 less than 5 in the ones place. What number am I?

d) I have 1 in the hundreds place, 4 in the tens place and 2 in the ones place. What number am I?

e) I have 1 in the hundreds place, 0 in the tens place and 0 in the ones place. What number am I?

## Module 11 Practice Lesson 2

24. Look at the following diagram and find the number.

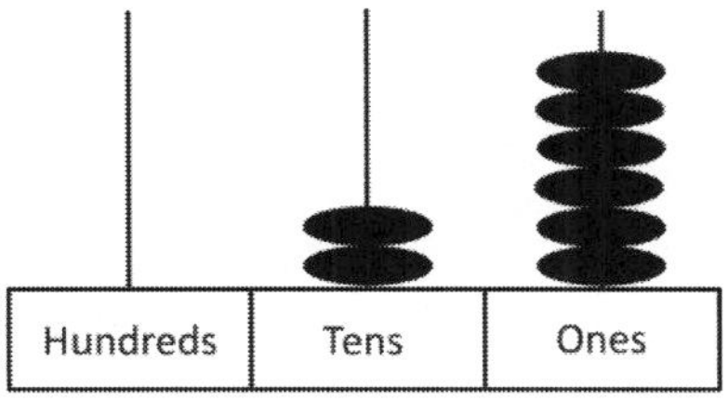


______________________

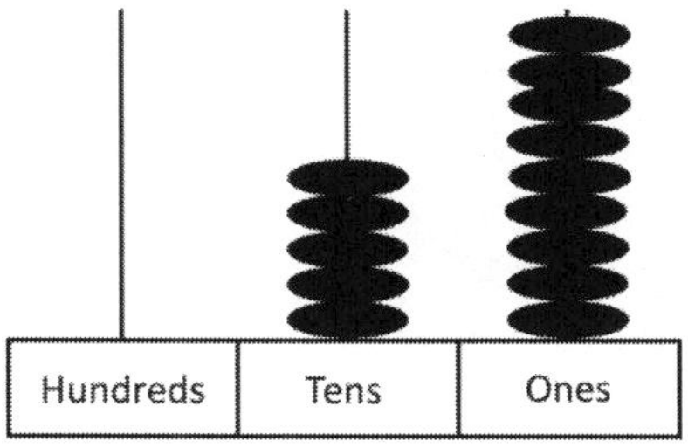


______________________

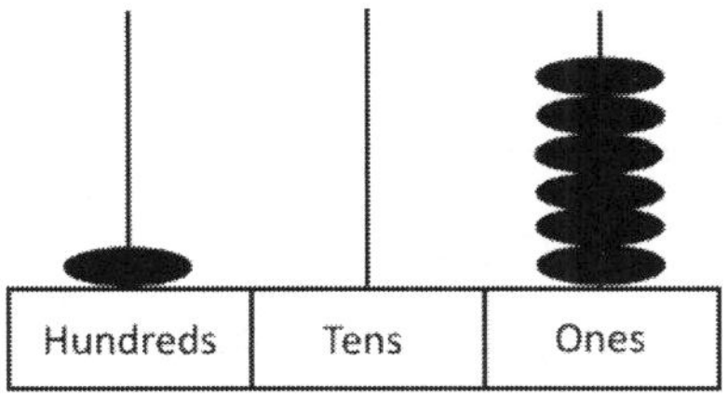


______________________

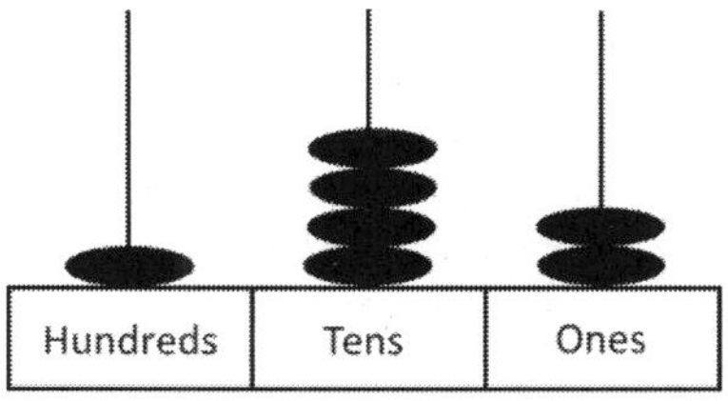


______________________

25. Put the correct number into the boxes.

| | | | | | | | |
|---|---|---|---|---|---|---|---|
| 14 | 26 | 34 | 84 | 56 | 74 | 96 | 94 |
| 104 | 116 | 4 | 16 | 36 | 76 | 54 | 66 |

4 in ones place

6 in ones place

Number larger than 50

Number smaller than 50

26. Try to find the number of cubes without counting them one by one.

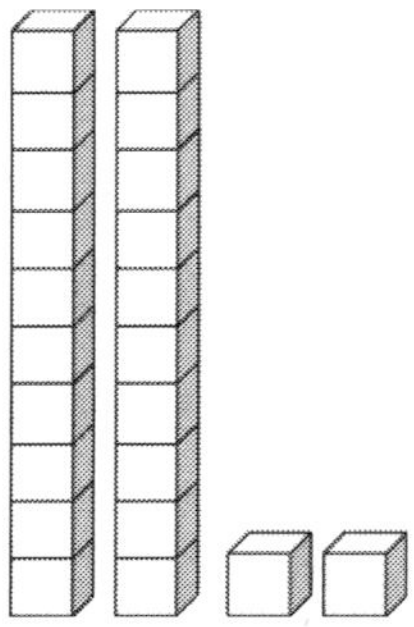

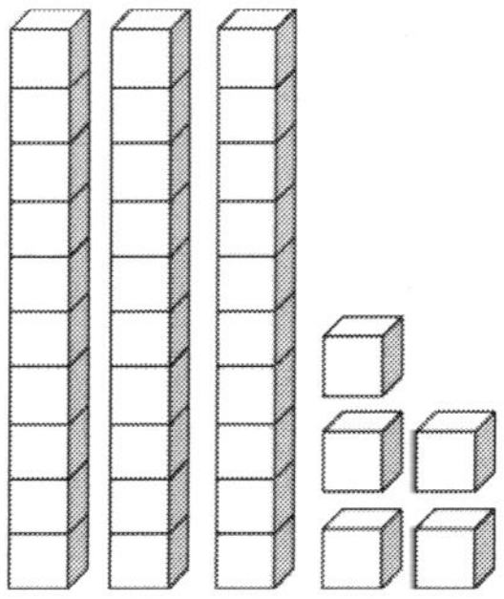

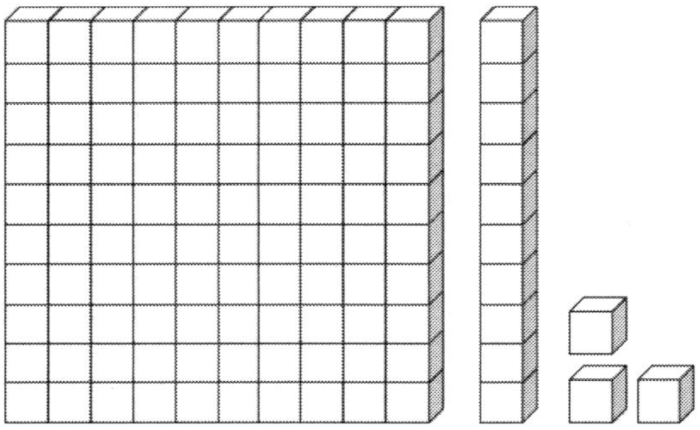

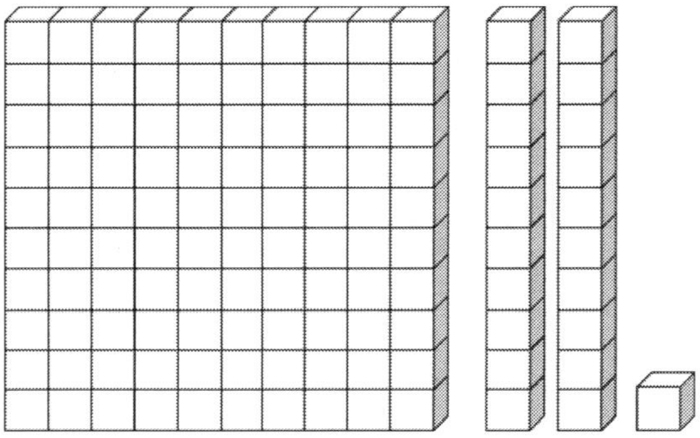

27. Complete the table below and answer the following questions.

| | | | | | | | | | |
|---|---|---|---|---|---|---|---|---|---|
| 1 | | 3 | | | | | 8 | | 10 |
| | 12 | | 14 | | | 17 | | 19 | |
| 21 | | 23 | | 25 | 26 | | 28 | | 30 |
| | 32 | | 34 | | | 37 | | 39 | |
| | | 43 | | 45 | 46 | | 48 | | |
| | | 53 | | 55 | 56 | | 58 | | |
| | 62 | | 64 | | | 67 | | 69 | |
| 71 | | 73 | | 75 | 76 | | 78 | | 80 |
| | 82 | | 84 | | | 87 | | 89 | |
| 91 | | 93 | | | | | 98 | | 100 |

a) Colour the numbers with 5 in the ones place in red.

b) Colour the numbers with 8 in tens place in yellow.

c) What is the 8th number in the second row?

d) What is the 6th number in the 8th row?

## Module 11 Practice Lesson 3

31. Find how many white squares and black squares are there without counting them one by one.

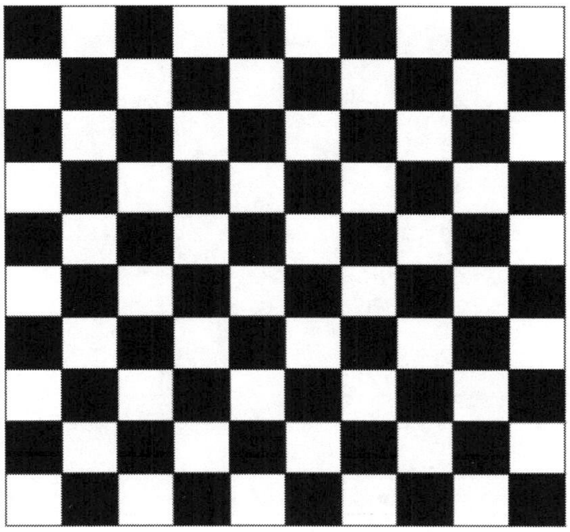

a) There are ____________ black cubes

b) There are ____________ white cubes

c) There are ____________ cubes in total

32. Compare the following numbers by using “>”, “<” and “=”.

| Hundreds | Tens | Ones |
|---|---|---|
| 0 | 9 | 4 |

○

| Hundreds | Tens | Ones |
|---|---|---|
| 0 | 7 | 5 |

| Hundreds | Tens | Ones |
|---|---|---|
| 0 | 7 | 4 |

○

| Hundreds | Tens | Ones |
|---|---|---|
| 0 | 7 | 5 |

| Hundreds | Tens | Ones |
|---|---|---|
| 1 | 1 | 3 |

○

| Hundreds | Tens | Ones |
|---|---|---|
| 1 | 0 | 4 |

| Hundreds | Tens | Ones |
|---|---|---|
| 1 | 2 | 7 |

○

| Hundreds | Tens | Ones |
|---|---|---|
| 1 | 2 | 0 |

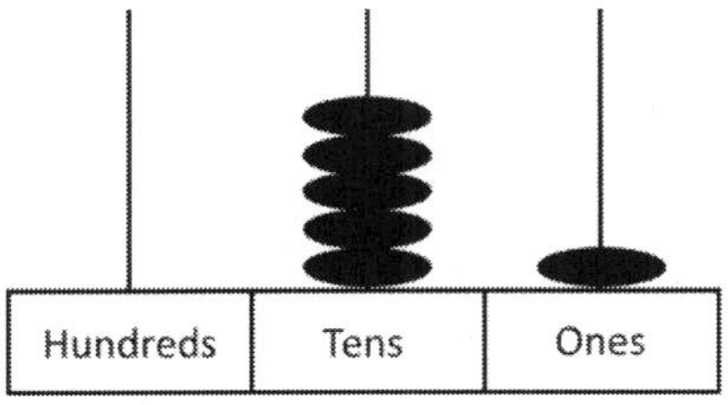


○

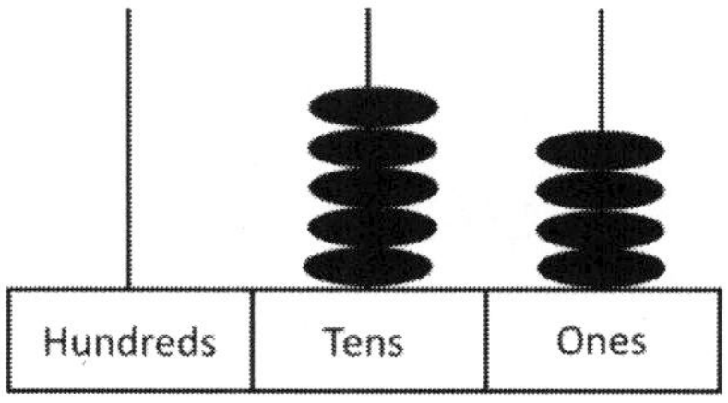


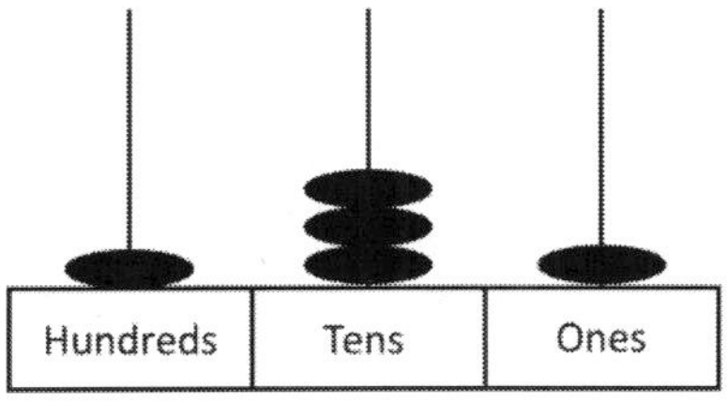


○

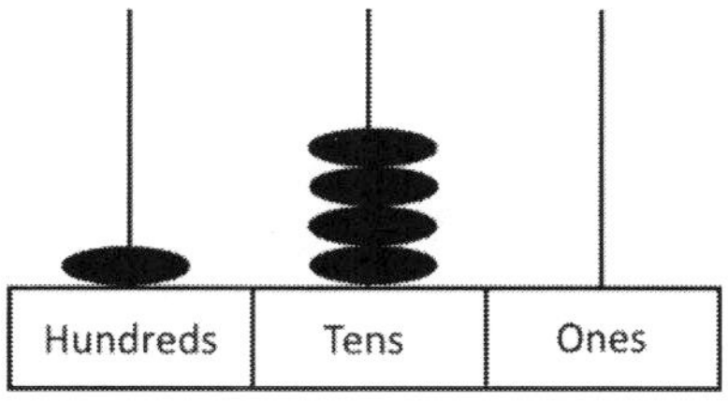

33. Find the number of marbles for each student from their conversation.

| | | |
|---|---|---|
| 20 marbles | 34 marbles | 74 marbles |
| | | |

34. Compare the following numbers and put them into the missing boxes.

| 57 | 21 | 43 | 85 | 77 |
|---|---|---|---|---|

☐ > ☐ > ☐ > ☐ > ☐

35. Write down five numbers that follow each rule below.

| | |
|---|---|
| Numbers larger than 48 | Numbers smaller than 35 |
| Numbers have 7 in the ones place | Numbers have 2 in the tens place |
| Numbers have 3-digit | Numbers have include 0 |

## Module 11 Practice Lesson 4

25. Complete the number sentence for the following diagrams.

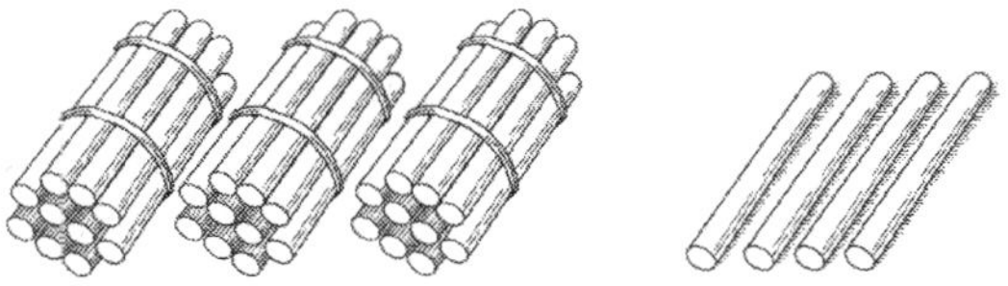

□ + 4 = □

□ − 4 = □

□ − □ = 4

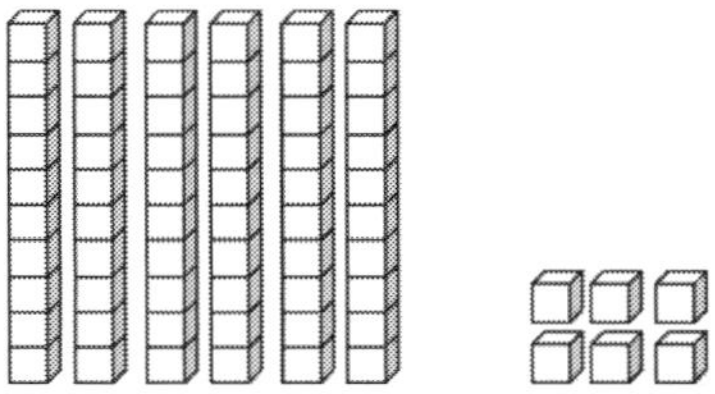

□ + 6 = □

□ − 6 = □

□ − □ = 6

26. Calculate the following number sentences.

| | | | |
|---|---|---|---|
| 10 + 8 = | 20 + 9 = | 30 + 7 = | 50 + 2 = |
| 14 + 5 = | 13 + 5 = | 24 + 3 = | 37 + 2 = |
| 46 − 6 = | 58 − 3 = | 74 − 3 = | 89 − 5 = |
| 53 − 50 = | 74 − 70 = | 82 − 80 = | 18 − 10 = |
| 47 − 20 = | 52 − 30 = | 74 − 50 = | 94 − 20 = |

27. Draw the line to help the dog find their home.

| | | | | | |
|---|---|---|---|---|---|
| 28 − 8 | 45 + 3 | 74 − 40 | 71 + 5 | 62 + 4 | 87 − 5 |

| | | | | | |
|---|---|---|---|---|---|
| 76 | 66 | 82 | 20 | 34 | 48 |

28. Jessica is learning Chinese every day. She already learned 74 Chinese characters today. And she still need to remember another 5 characters. How many characters she need to remember in total?

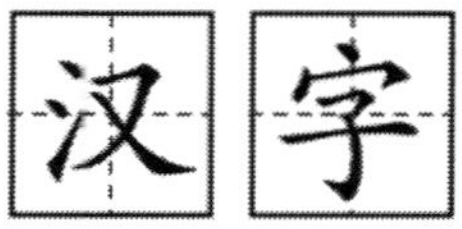


29. George is making some handcrafts for the Christmas party. He needs to make 75 handcrafts in total. He already made 60 handcrafts. How many more does he need to make?

30. Mr. Chen bought some candies for his 75 students. If each student can have 1 candy, then there are 4 candies left. How many candies did Mr. Chen buy in total?

31. Jimmy collected 87 bottles. If each box can hold 60 bottles, how many bottles are not in the box?

# Module 12 找规律 Find the pattern

❖ **跳数的认识 Skip counting**

Skip counting is counting by a number that is more than 1. Here are some examples for skip counting.

Skip counting by 2

| | | | | | | | | | |
|---|---|---|---|---|---|---|---|---|---|
| 1 | **2** | 3 | **4** | 5 | **6** | 7 | **8** | 9 | **10** |
| 11 | **12** | 13 | **14** | 15 | **16** | 17 | **18** | 19 | **20** |
| 21 | **22** | 23 | **24** | 25 | **26** | 27 | **28** | 29 | **30** |
| 31 | **32** | 33 | **34** | 35 | **36** | 37 | **38** | 39 | **40** |
| 41 | **42** | 43 | **44** | 45 | **46** | 47 | **48** | 49 | **50** |
| 51 | **52** | 53 | **54** | 55 | **56** | 57 | **58** | 59 | **60** |
| 61 | **62** | 63 | **64** | 65 | **66** | 67 | **68** | 69 | **70** |
| 71 | **72** | 73 | **74** | 75 | **76** | 77 | **78** | 79 | **80** |
| 81 | **82** | 83 | **84** | 85 | **86** | 87 | **88** | 89 | **90** |
| 91 | **92** | 93 | **94** | 95 | **96** | 97 | **98** | 99 | **100** |

Skip counting by 3

| | | | | | | | | | |
|---|---|---|---|---|---|---|---|---|---|
| 1 | 2 | **3** | 4 | 5 | **6** | 7 | 8 | **9** | 10 |
| 11 | **12** | 13 | 14 | **15** | 16 | 17 | **18** | 19 | 20 |
| **21** | 22 | 23 | **24** | 25 | 26 | **27** | 28 | 29 | **30** |
| 31 | 32 | **33** | 34 | 35 | **36** | 37 | 38 | **39** | 40 |
| 41 | **42** | 43 | 44 | **45** | 46 | 47 | **48** | 49 | 50 |
| **51** | 52 | 53 | **54** | 55 | 56 | **57** | 58 | 59 | **60** |
| 61 | 62 | **63** | 64 | 65 | **66** | 67 | 68 | **69** | 70 |
| 71 | **72** | 73 | 74 | **75** | 76 | 77 | **78** | 79 | 80 |
| **81** | 82 | 83 | **84** | 85 | 86 | **87** | 88 | 89 | **90** |
| 91 | 92 | **93** | 94 | 95 | **96** | 97 | 98 | **99** | 100 |

Skip counting by 5

| | | | | | | | | | |
|---|---|---|---|---|---|---|---|---|---|
| 1 | 2 | 3 | 4 | **5** | 6 | 7 | 8 | 9 | **10** |
| 11 | 12 | 13 | 14 | **15** | 16 | 17 | 18 | 19 | **20** |
| 21 | 22 | 23 | 24 | **25** | 26 | 27 | 28 | 29 | **30** |
| 31 | 32 | 33 | 34 | **35** | 36 | 37 | 38 | 39 | **40** |
| 41 | 42 | 43 | 44 | **45** | 46 | 47 | 48 | 49 | **50** |
| 51 | 52 | 53 | 54 | **55** | 56 | 57 | 58 | 59 | **60** |
| 61 | 62 | 63 | 64 | **65** | 66 | 67 | 68 | 69 | **70** |
| 71 | 72 | 73 | 74 | **75** | 76 | 77 | 78 | 79 | **80** |
| 81 | 82 | 83 | 84 | **85** | 86 | 87 | 88 | 89 | **90** |
| 91 | 92 | 93 | 94 | **95** | 96 | 97 | 98 | 99 | **100** |

Skip counting by 10

| | | | | | | | | | |
|---|---|---|---|---|---|---|---|---|---|
| 1 | 2 | 3 | 4 | 5 | 6 | 7 | 8 | 9 | **10** |
| 11 | 12 | 13 | 14 | 15 | 16 | 17 | 18 | 19 | **20** |
| 21 | 22 | 23 | 24 | 25 | 26 | 27 | 28 | 29 | **30** |
| 31 | 32 | 33 | 34 | 35 | 36 | 37 | 38 | 39 | **40** |
| 41 | 42 | 43 | 44 | 45 | 46 | 47 | 48 | 49 | **50** |
| 51 | 52 | 53 | 54 | 55 | 56 | 57 | 58 | 59 | **60** |
| 61 | 62 | 63 | 64 | 65 | 66 | 67 | 68 | 69 | **70** |
| 71 | 72 | 73 | 74 | 75 | 76 | 77 | 78 | 79 | **80** |
| 81 | 82 | 83 | 84 | 85 | 86 | 87 | 88 | 89 | **90** |
| 91 | 92 | 93 | 94 | 95 | 96 | 97 | 98 | 99 | **100** |

We can do skip counting both forwards (numbers become bigger) and backwards (numbers become smaller).

Skip counting helps you count many things quickly, so that you won't need to count items one by one.

Skip counting also helps us to learn the multiplication tables in next year's learning.

## 图像规律 Geometric patterns

Symbols and shapes can also form patterns.

For example:

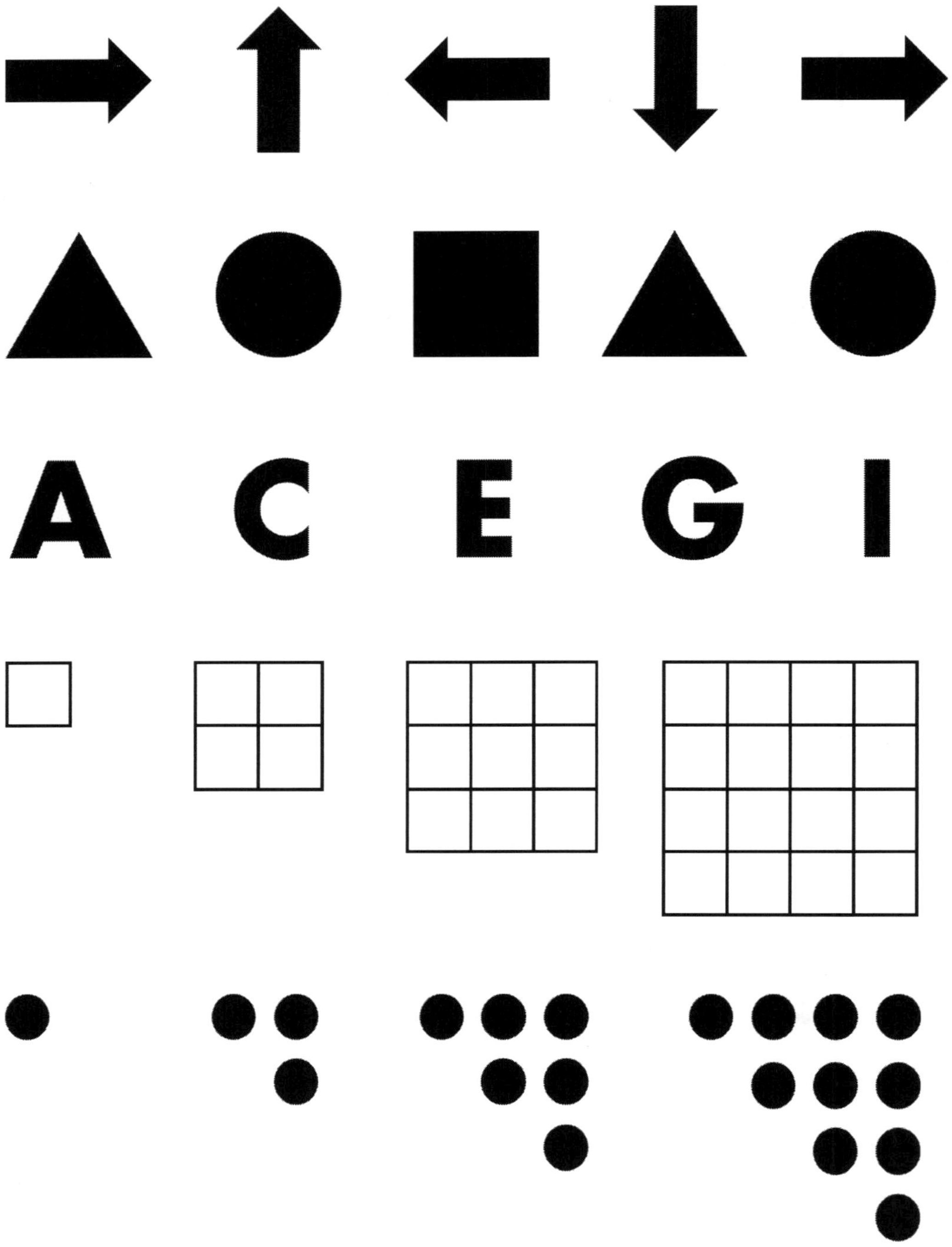

## Module 12 Practice Lesson 1

31. Count forwards by 2's.

| | | | | | | | | | |
|---|---|---|---|---|---|---|---|---|---|
| 2 | 4 | | | | | | | | |

| | | | | | | | | | |
|---|---|---|---|---|---|---|---|---|---|
| 20 | 22 | 24 | | | | | | | |

| | | | | | | | | | |
|---|---|---|---|---|---|---|---|---|---|
| 48 | 50 | | | | | | | | |

32. Count forwards by 3's.

| | | | | | | | | | |
|---|---|---|---|---|---|---|---|---|---|
| 3 | 6 | 9 | | | | | | | |

| | | | | | | | | | |
|---|---|---|---|---|---|---|---|---|---|
| 15 | 18 | 21 | | | | | | | |

| | | | | | | | | | |
|---|---|---|---|---|---|---|---|---|---|
| 30 | 33 | 36 | | | | | | | |

33. Count forwards by 5's.

| | | | | | | | | | |
|---|---|---|---|---|---|---|---|---|---|
| 5 | 10 | 15 | | | | | | | |

| | | | | | | | | | |
|---|---|---|---|---|---|---|---|---|---|
| 30 | 35 | | | | | | | | |

| | | | | | | | | | |
|---|---|---|---|---|---|---|---|---|---|
| 50 | | | | | | | | | |

34. Fill in the missing number in each pattern.

| | | | | | | | | | |
|---|---|---|---|---|---|---|---|---|---|
| 3 | 5 | | 9 | | 13 | | 17 | | 21 |

| | | | | | | | | | |
|---|---|---|---|---|---|---|---|---|---|
| 10 | | 30 | | 50 | | 70 | | 90 | |

| | | | | | | | | | |
|---|---|---|---|---|---|---|---|---|---|
| 0 | | 10 | 15 | | | | | 40 | |

| | | | | | | | | | |
|---|---|---|---|---|---|---|---|---|---|
| 10 | 13 | | | 22 | | 28 | | | 37 |

35. Draw the next three shapes in each pattern.

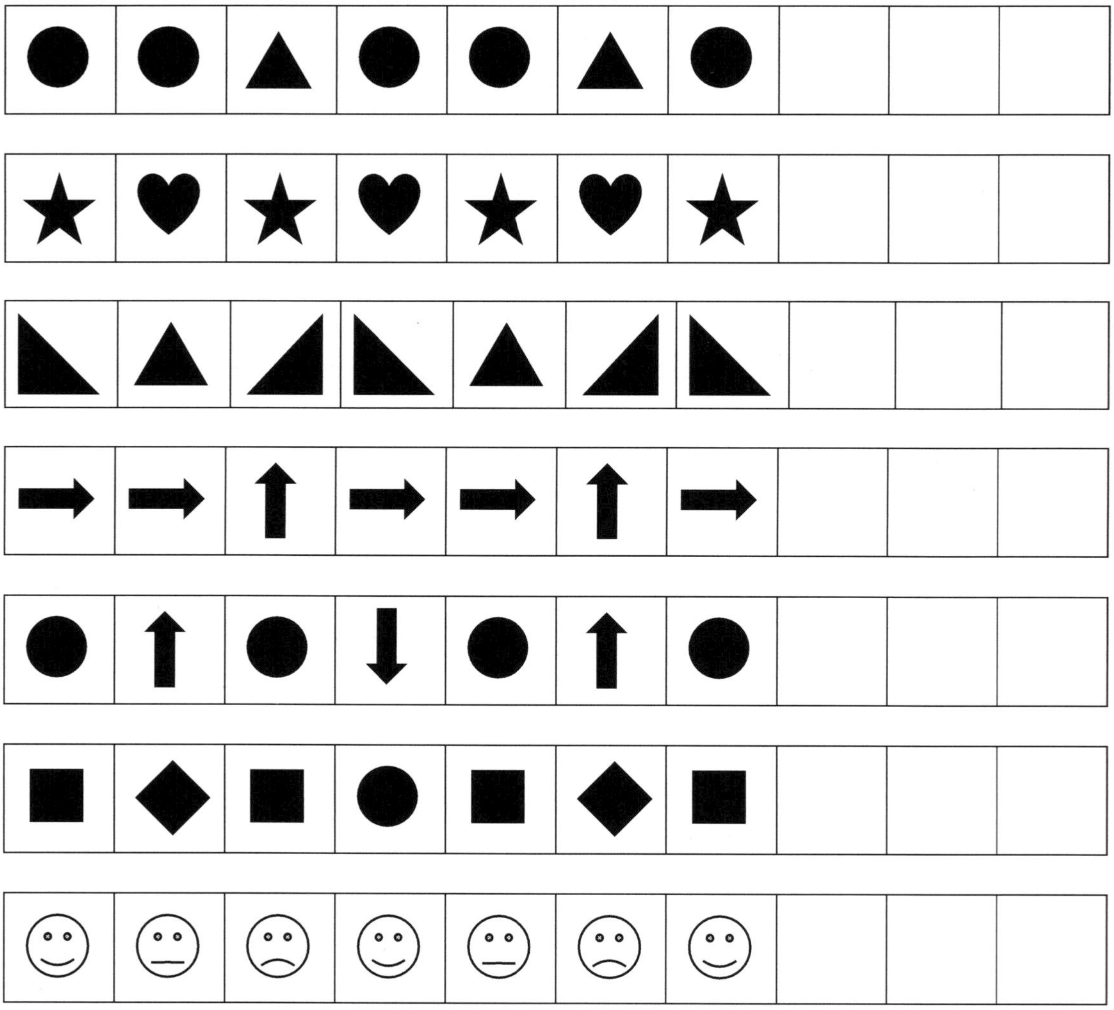

## Module 12 Practice Lesson 2

1. Count backwards by 2's.

| | | | | | | | | | |
|---|---|---|---|---|---|---|---|---|---|
| 22 | 20 | 18 | | | | | | | |

| | | | | | | | | | |
|---|---|---|---|---|---|---|---|---|---|
| 74 | 72 | | | | | | | | |

| | | | | | | | | | |
|---|---|---|---|---|---|---|---|---|---|
| 60 | | | | | | | | | |

2. Count backwards by 3's.

| | | | | | | | | | |
|---|---|---|---|---|---|---|---|---|---|
| 36 | 33 | 30 | | | | | | | |

| | | | | | | | | | |
|---|---|---|---|---|---|---|---|---|---|
| 45 | 42 | | | | | | | | |

| | | | | | | | | | |
|---|---|---|---|---|---|---|---|---|---|
| 90 | | | | | | | | | |

3. Count backwards by 5's.

| | | | | | | | | | |
|---|---|---|---|---|---|---|---|---|---|
| 70 | 65 | 60 | | | | | | | |

| | | | | | | | | | |
|---|---|---|---|---|---|---|---|---|---|
| 85 | 80 | | | | | | | | |

| | | | | | | | | | |
|---|---|---|---|---|---|---|---|---|---|
| 100 | | | | | | | | | |

4. Find the missing numbers.

| | | | | | | | | | |
|---|---|---|---|---|---|---|---|---|---|
| 42 | 40 | 38 | | 34 | | | 28 | | 24 |

| | | | | | | | | | |
|---|---|---|---|---|---|---|---|---|---|
| 80 | | | 74 | 72 | | | 66 | | 62 |

| | | | | | | | | | |
|---|---|---|---|---|---|---|---|---|---|
| | | 70 | 65 | | 55 | | 45 | | |

5. Find the pattern and draw the next shape in the grid.

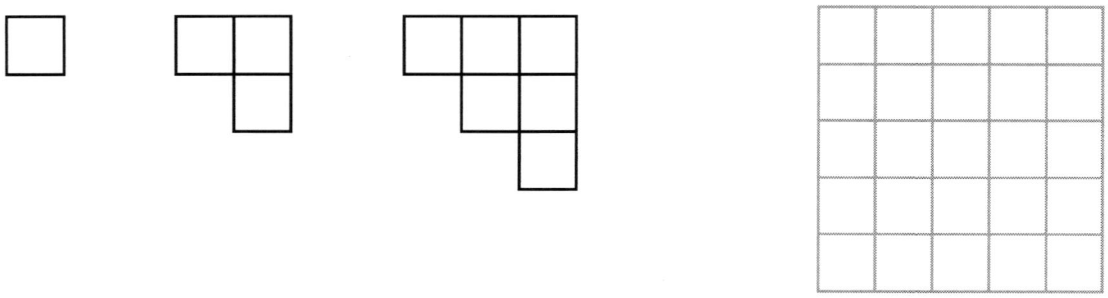

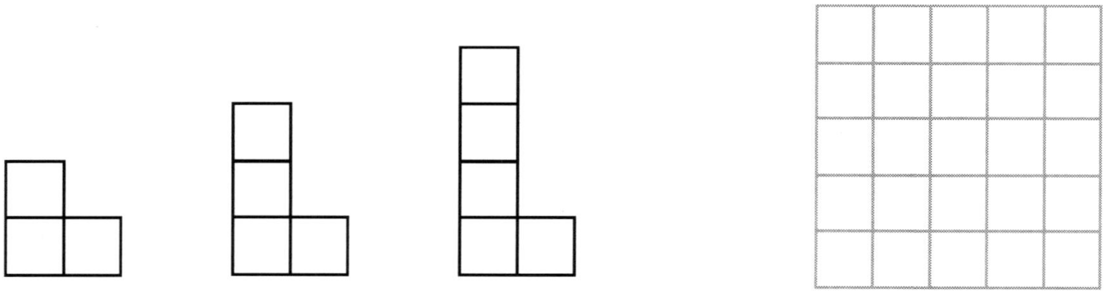

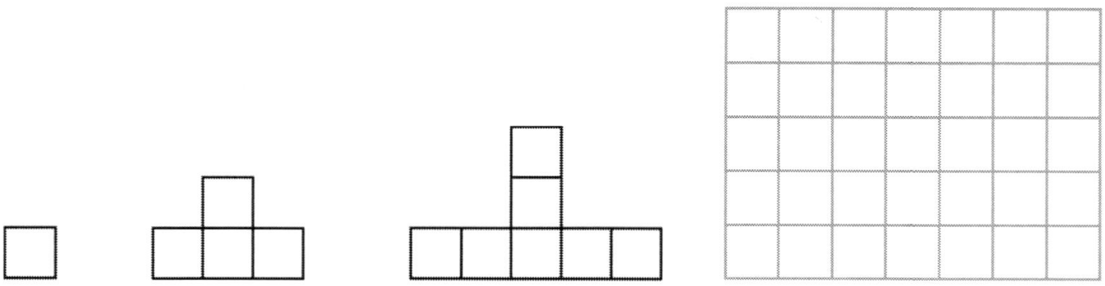

## Module 12 Practice Lesson 3

36. Using the hundreds chart below, skip counting by 2s starting from 0. Colour the number than you land on each time. Answer the following questions.

| | | | | | | | | | |
|---|---|---|---|---|---|---|---|---|---|
| 1 | 2 | 3 | 4 | 5 | 6 | 7 | 8 | 9 | 10 |
| 11 | 12 | 13 | 14 | 15 | 16 | 17 | 18 | 19 | 20 |
| 21 | 22 | 23 | 24 | 25 | 26 | 27 | 28 | 29 | 30 |
| 31 | 32 | 33 | 34 | 35 | 36 | 37 | 38 | 39 | 40 |
| 41 | 42 | 43 | 44 | 45 | 46 | 47 | 48 | 49 | 50 |
| 51 | 52 | 53 | 54 | 55 | 56 | 57 | 58 | 59 | 60 |
| 61 | 62 | 63 | 64 | 65 | 66 | 67 | 68 | 69 | 70 |
| 71 | 72 | 73 | 74 | 75 | 76 | 77 | 78 | 79 | 80 |
| 81 | 82 | 83 | 84 | 85 | 86 | 87 | 88 | 89 | 90 |
| 91 | 92 | 93 | 94 | 95 | 96 | 97 | 98 | 99 | 100 |

a) What is the smallest number you have shaded?

b) What is the largest number you have shaded?

c) What pattern on the ones place when skip counting by 2?

37. Find the missing numbers.

| | | | | | | | | | |
|---|---|---|---|---|---|---|---|---|---|
| 12 | | 32 | | | 62 | 72 | | | 102 |

| | | | | | | | | | |
|---|---|---|---|---|---|---|---|---|---|
| 14 | | 24 | | | 39 | | 49 | 54 | |

| | | | | | | | | | |
|---|---|---|---|---|---|---|---|---|---|
| | | | 33 | | | 24 | | 18 | 15 |

| | | | | | | | | | |
|---|---|---|---|---|---|---|---|---|---|
| 53 | | 49 | | | 43 | 41 | | | 35 |

| | | | | | | | | | |
|---|---|---|---|---|---|---|---|---|---|
| 74 | 77 | | | 86 | | | 95 | | 101 |

| | | | | | | | | | |
|---|---|---|---|---|---|---|---|---|---|
| 43 | | 53 | 58 | | | 73 | | | 88 |

38. Observe the shape and find each pattern.

a)

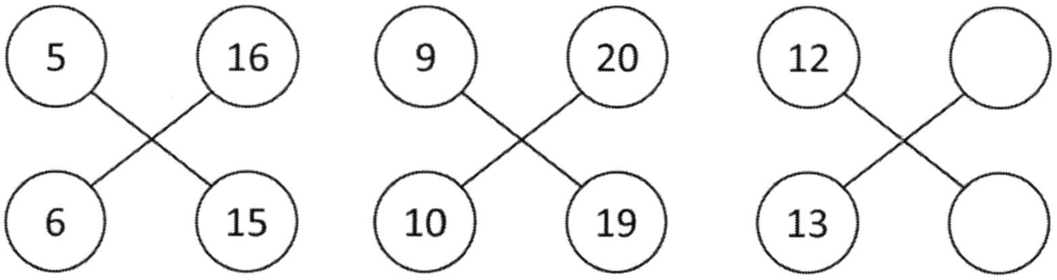


b)

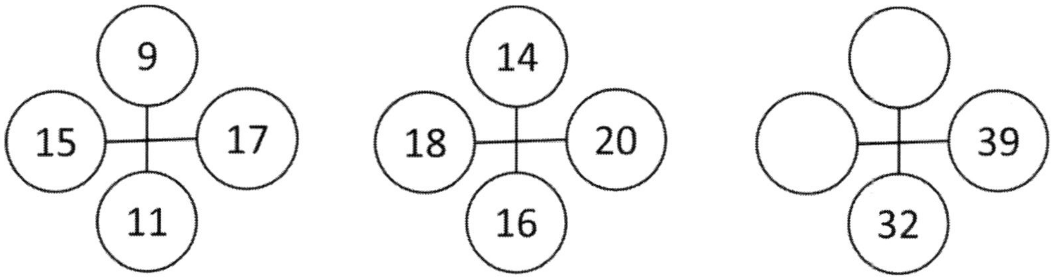


c)

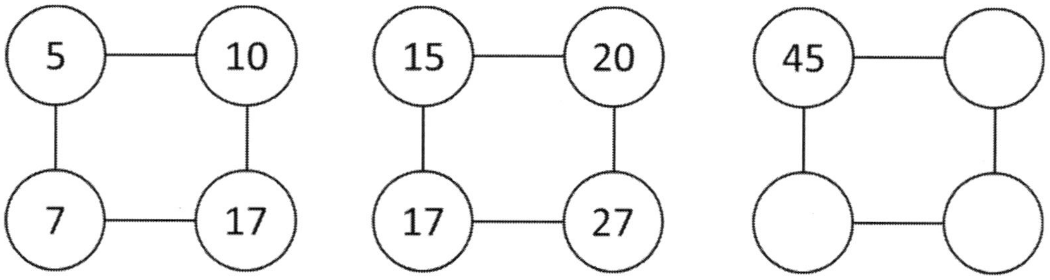


39. Serena has some match sticks. She used 4 matches to make a square. Then she added more matches to the right to form the shape below.

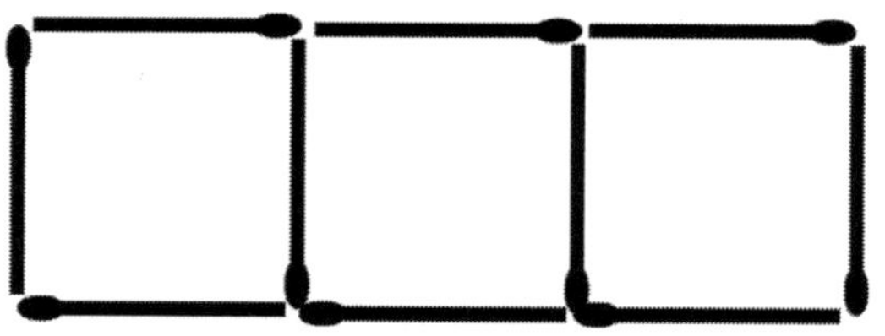

a) How many match sticks are there when she made 3 connected squares?

b) Can you draw the matches if she wants to add one more square on the shape?

## Module 12 Practice Lesson 4

1. Using the hundreds chart below, skip counting by 5s starting from 0. Colour the number than you land on each time. Answer the following questions.

| | | | | | | | | | |
|---|---|---|---|---|---|---|---|---|---|
| 1 | 2 | 3 | 4 | 5 | 6 | 7 | 8 | 9 | 10 |
| 11 | 12 | 13 | 14 | 15 | 16 | 17 | 18 | 19 | 20 |
| 21 | 22 | 23 | 24 | 25 | 26 | 27 | 28 | 29 | 30 |
| 31 | 32 | 33 | 34 | 35 | 36 | 37 | 38 | 39 | 40 |
| 41 | 42 | 43 | 44 | 45 | 46 | 47 | 48 | 49 | 50 |
| 51 | 52 | 53 | 54 | 55 | 56 | 57 | 58 | 59 | 60 |
| 61 | 62 | 63 | 64 | 65 | 66 | 67 | 68 | 69 | 70 |
| 71 | 72 | 73 | 74 | 75 | 76 | 77 | 78 | 79 | 80 |
| 81 | 82 | 83 | 84 | 85 | 86 | 87 | 88 | 89 | 90 |
| 91 | 92 | 93 | 94 | 95 | 96 | 97 | 98 | 99 | 100 |

a) What is the smallest number you have shaded?

b) What is the largest number you have shaded?

c) What pattern on the ones place when skip counting by 5?

32. Observe the shape and find each pattern.

a)

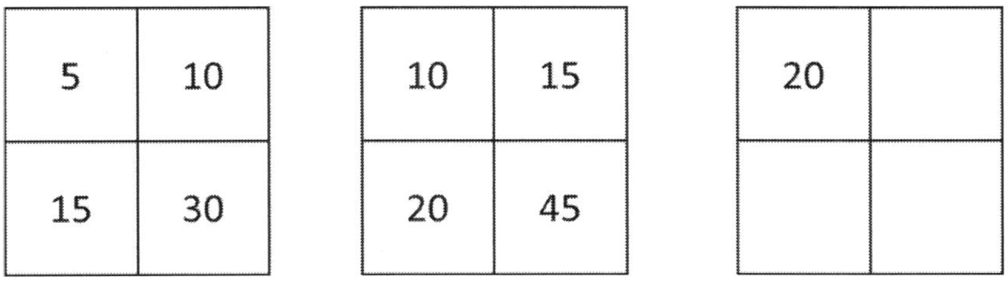


b)

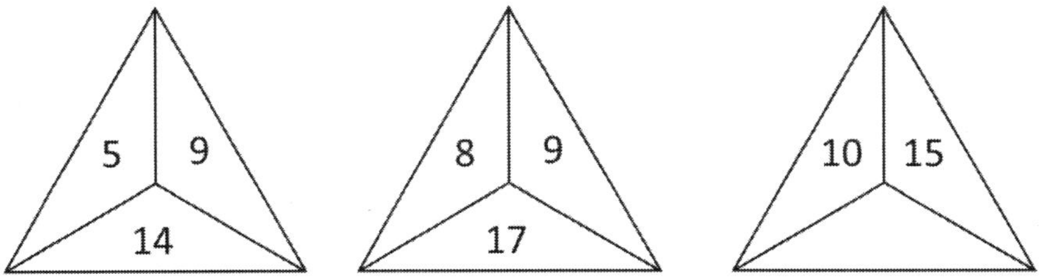


c)

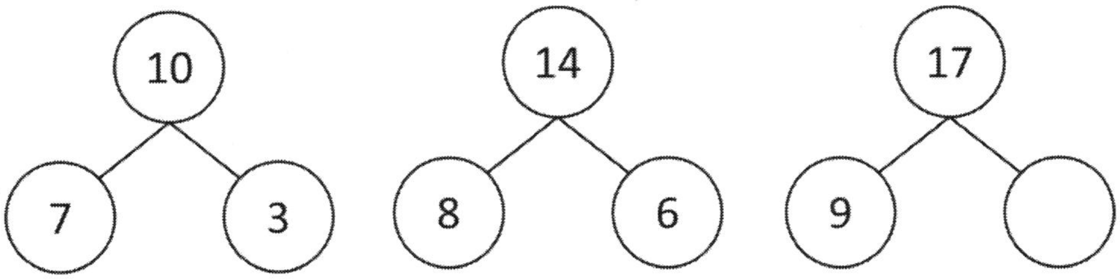

33. Amy has some match sticks. She used 3 matches to make a triangle. Then she added more matches to the right to form the shape below.

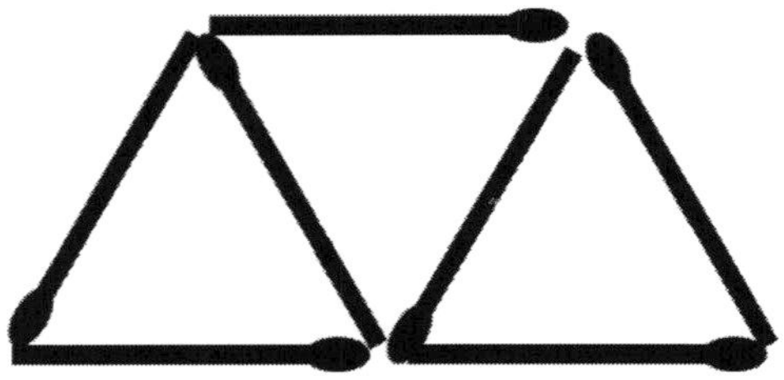

a) How many matches did she use to make 3 connected triangles?

b) If she add 1 more triangle on the shape above, how many more matches does she need?

c) Draw the next shape and count how many matches she used in total.

34. Nami, the dog needs to be trained to catch the ball every day. She can catch 2 balls on the first day. 4 balls on the second day, and 6 balls on the third day.
How many balls can she catch on the 6th day? And how about on the 10th day?

## Module 13 100 以内的计算 Calculation between 0 – 100

### ❖ 整十的加减 Adding and subtracting 10s number

➢ Example 1:

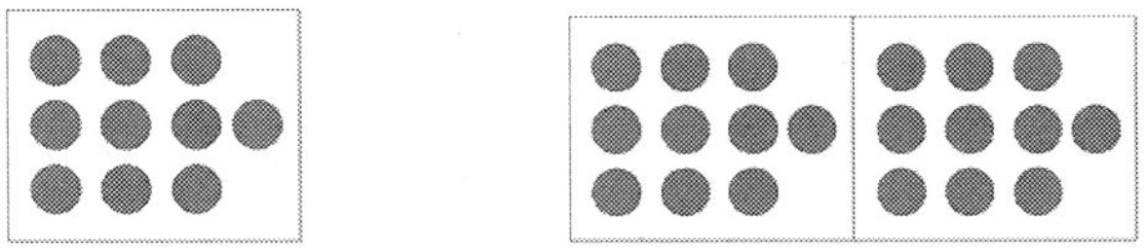

10 + 20 = 30

1 + 2

➢ Example 2:

$ 50 $ 70

50 + 70 = 120

5 + 7

➢ Example 3:

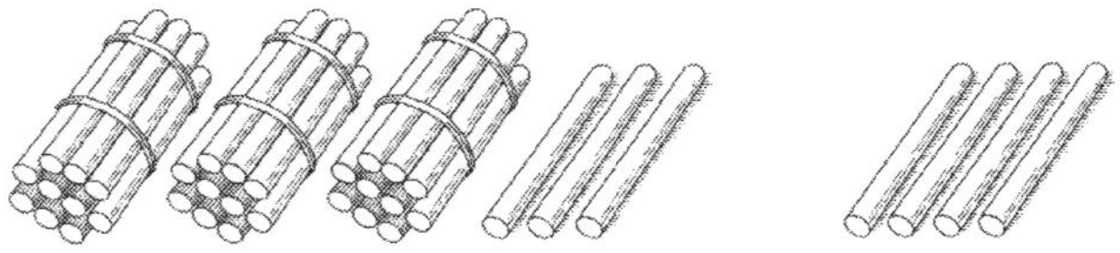

33 + 3 = 36

30 3

6

➢ Example 4:

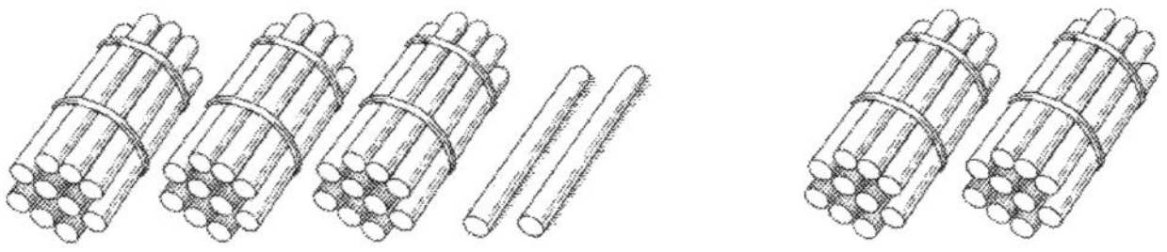

32 + 20 = 52

30 3

50

- **两位数加一位数** **2-digit number add 1-digit number**

  - Example 1:

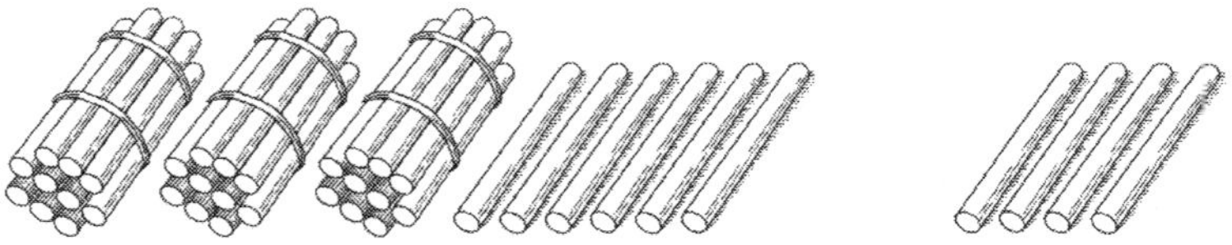

$$36 + 4 = 40$$

36 → 30, 6; 6 + 4 → 10

  - Example 2:

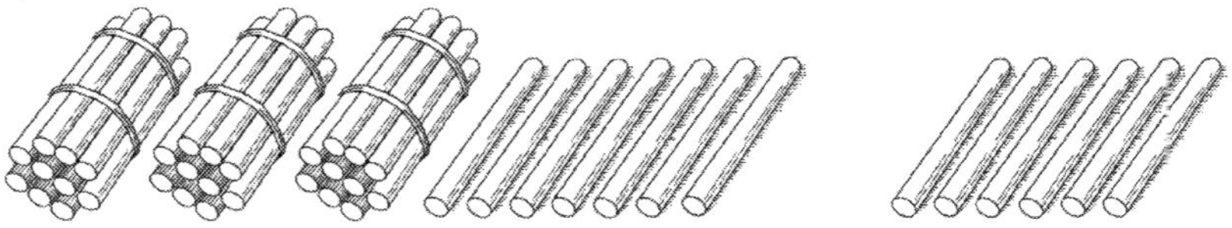

$$37 + 6 = 43$$

37 → 30, 7; 7 + 6 → 13

- **两位数减一位数** **2-digit number subtract 1-digit number**

  - Example 1:

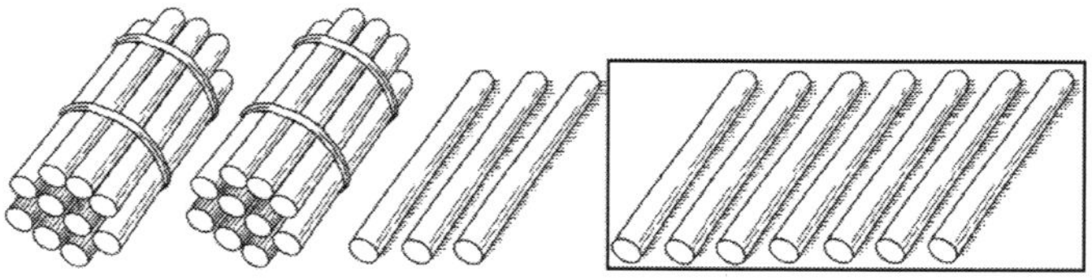

$$30 - 7 = 23$$

30 → 20, 10; 10 − 7 → 3

  - Example 2:

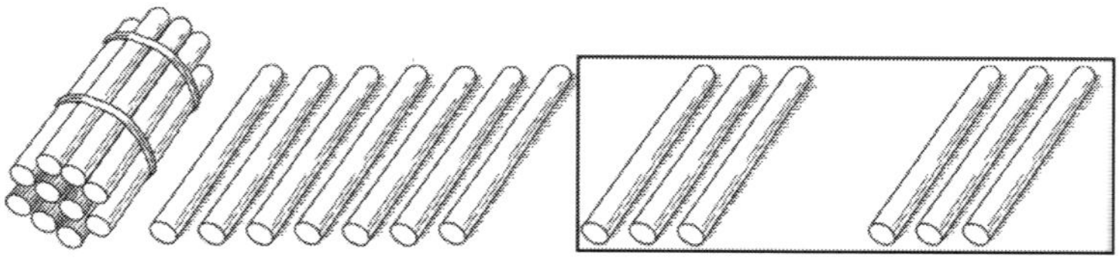

$$23 - 6 = 17$$

23 → 10, 13; 13 − 6 → 7

## Module 13 Practice Lesson 1

36. Complete the number sentences by looking at the following diagrams.

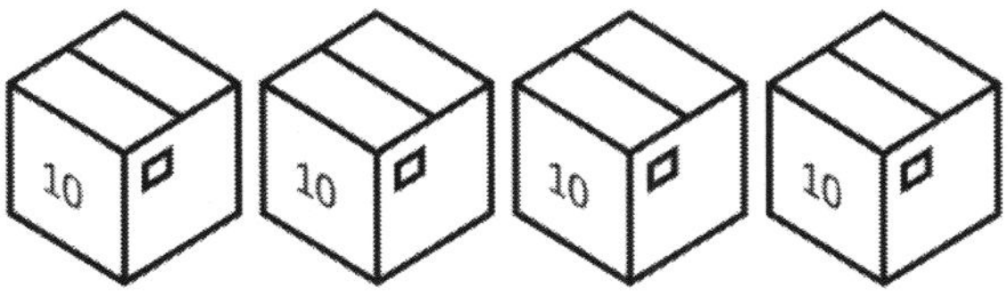


40 + 30 = ☐

☐ + ☐ = ☐

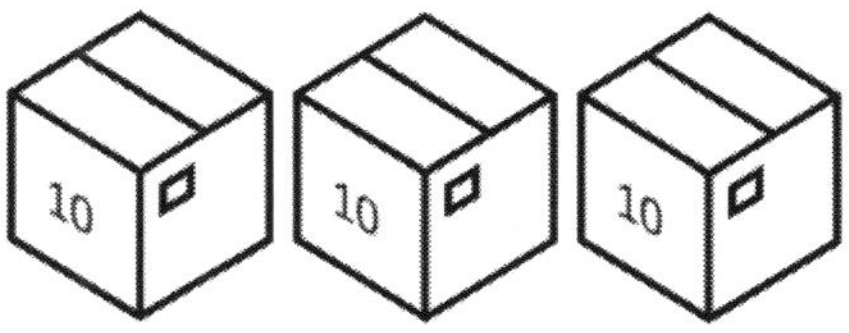


☐ − ☐ = ☐

☐ − ☐ = ☐

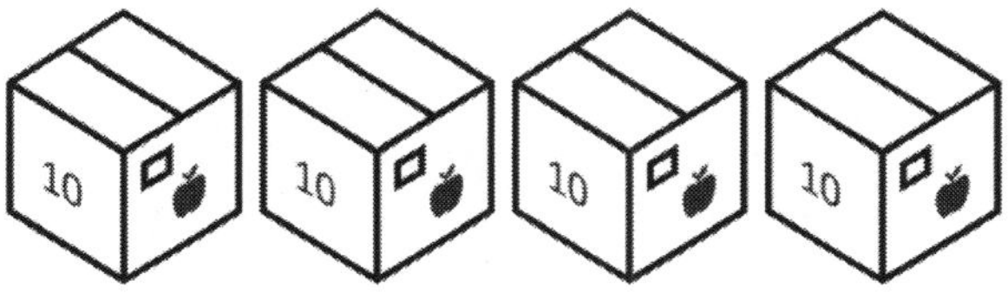


☐ + ☐ = ☐

☐ + ☐ = ☐

☐ − ☐ = ☐

☐ − ☐ = ☐

37. Calculate the following numbers for each group.

| |
|---|
| 5 + 4 = |
| 65 + 4 = |

| |
|---|
| 7 + 2 = |
| 67 + 2 = |

| |
|---|
| 10 + 40 = |
| 15 + 40 = |

| |
|---|
| 60 + 20 = |
| 67 + 20 = |

| |
|---|
| 20 + 5 = |
| 20 + 25 = |

| |
|---|
| 10 + 2 = |
| 10 + 32 = |

38. Find the answer for each question.

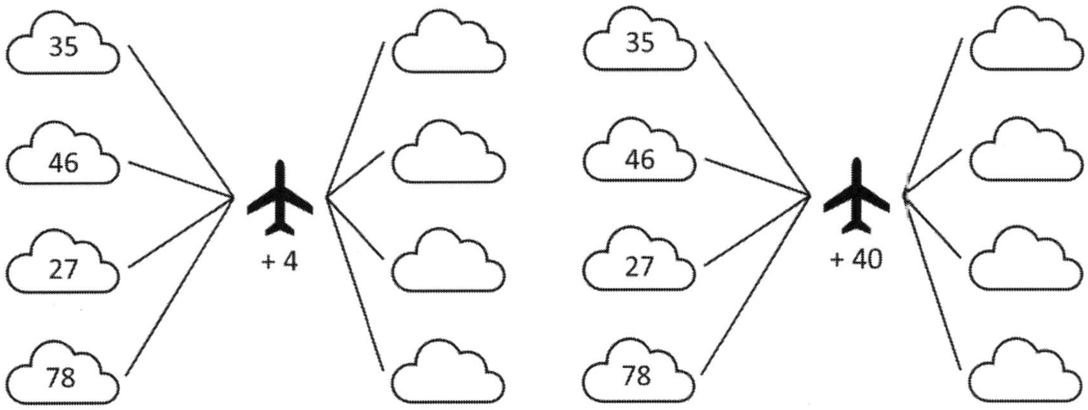


39. Use number 20, 30, 40, 50, 60 and 70 to complete the following number sentence.

☐ + ☐ = ☐ + ☐ = ☐ + ☐

☐ − ☐ = ☐ − ☐ = ☐ − ☐

40. Joe is shopping for his school open day. Here are some stationaries listed below.

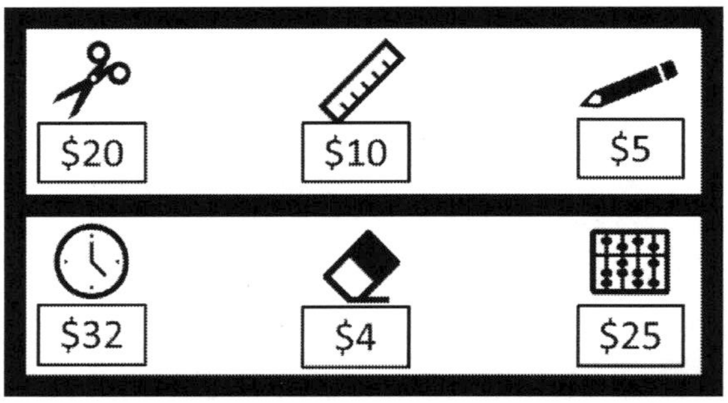


a) If he wants to buy an abacus and a ruler, how much money does he need to pay?

b) If he wants to buy a clock, a ruler and a rubber, how much money does he need to pay?

c) He gives the cashier $50 to buy a scissors and a ruler. How much change should he get back?

## Module 13 Practice Lesson 2

6. Group the sticks into "10" and complete the question.

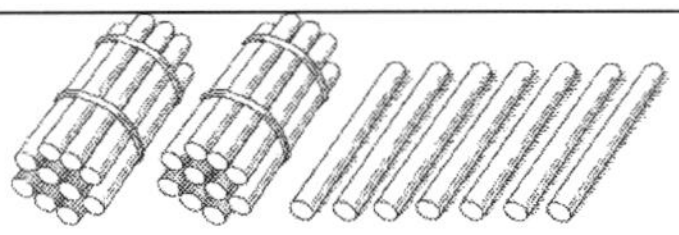

27 + 3 = □

First calculate:

7 + 3 = □

And then calculate :

□ + □ = □

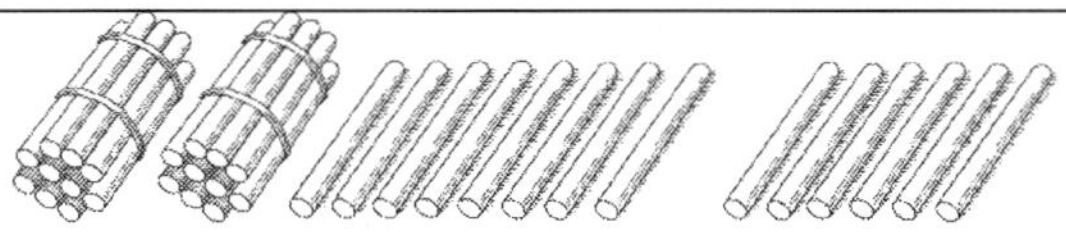

28 + 6 = □

First calculate:

8 + 6 = □

And then calculate :

□ + □ = □

7. Fill in the missing number and find the answer. The first question has been done for you.

| | | |
|---|---|---|
| 37 + 6 = 43 (30, 7; 7 + 6 = 13) | 45 + 7 = □ | 74 + 8 = □ |
| 47 + 6 = □ | 84 + 9 = □ | 46 + 5 = □ |

8. Calculate the following questions.

| | |
|---|---|
| 4 + 7 = | 5 + 8 = |
| 34 + 7 = | 75 + 8 = |
| 9 + 3 = | 7 + 5 = |
| 69 + 3 = | 7 + 65 = |

9. Donnie did the following questions. Did he do correct? If not, find the mistake and rewrite it.

| | |
|---|---|
| 45 + 6 = 41 | |
| 58 + 3 = 88 | |
| 78 + 7 = 95 | |
| 6 + 35 = 91 | |

10. Compare the following number sentences by using “>”, “<” or “=”.

34 + 7 ○ 35 + 4    74 + 6 ○ 75 + 7    17 + 7 ○ 19 + 5

74 + 5 ○ 68 + 9    35 + 7 ○ 34 + 8    44 + 5 ○ 45 + 6

45 + 6 ○ 47 + 4    75 + 6 ○ 72 + 7    75 + 9 ○ 77 + 4

11. Gina wants to buy some toys for her daughter.

| | | |
|---|---|---|
| $9 | $38 | $40 |

a) If Gina want to spend the least money, which two toys she can buy?

b) If Gina want to spend the most money, which two toys she can buy?

## Module 13 Practice Lesson 3

**40.** Complete the number sentence below.

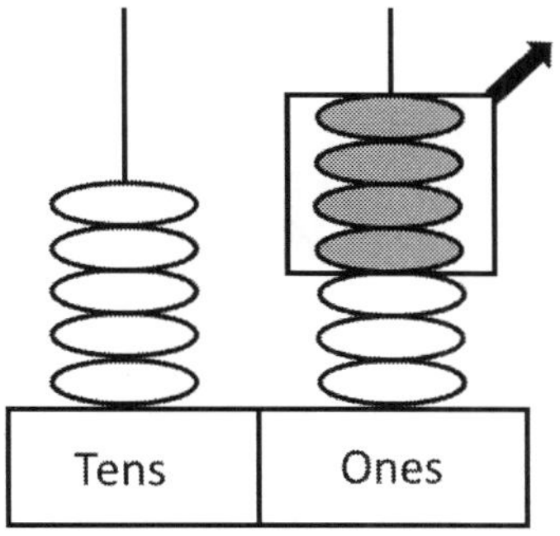


☐ − ☐ = ☐

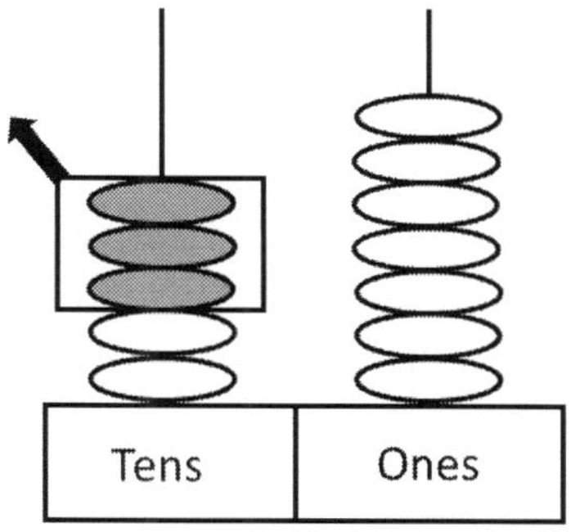


☐ − ☐ = ☐

**41.** Find the missing numbers in the following questions. The first question has been done for you.

59 − 6 = 53

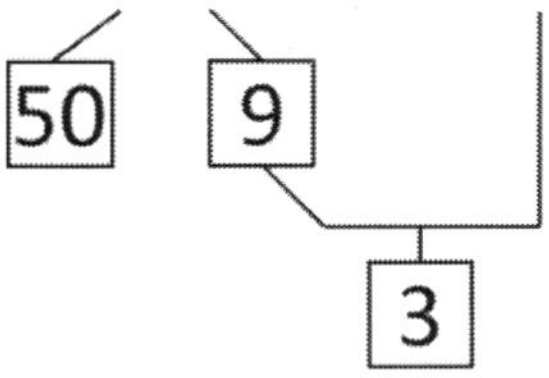


47 − 3 = ☐

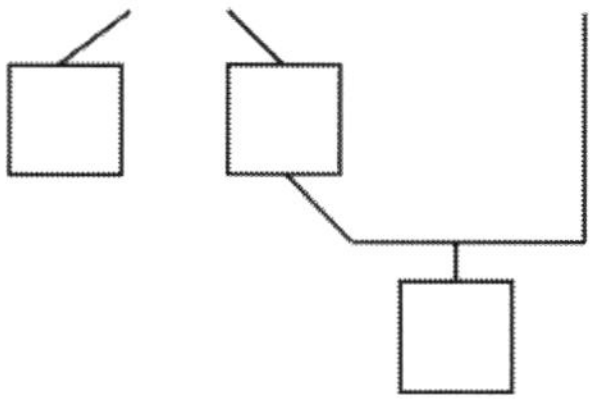

87 − 6 = ☐

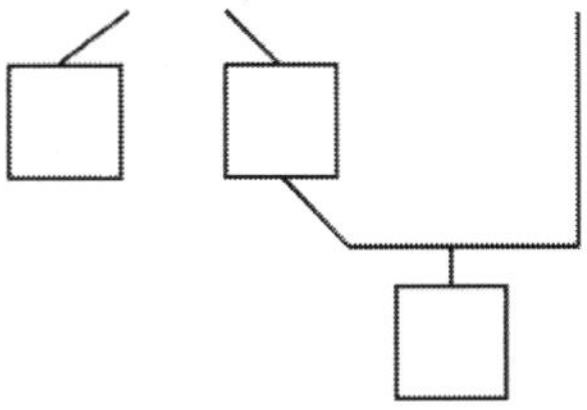

82 − 30 = ☐

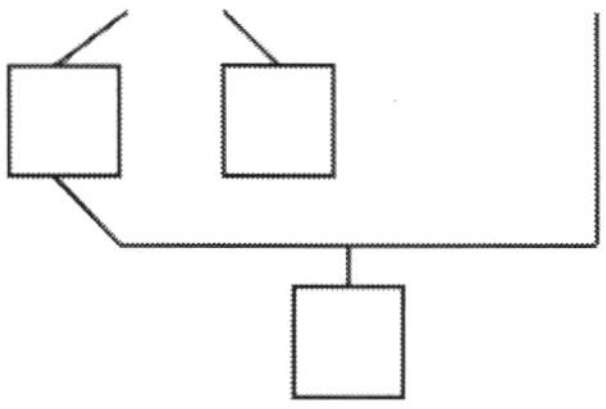

53 − 20 = ☐

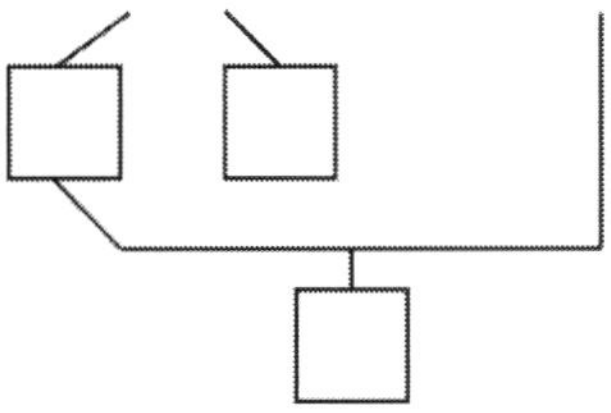

74 − 50 = ☐

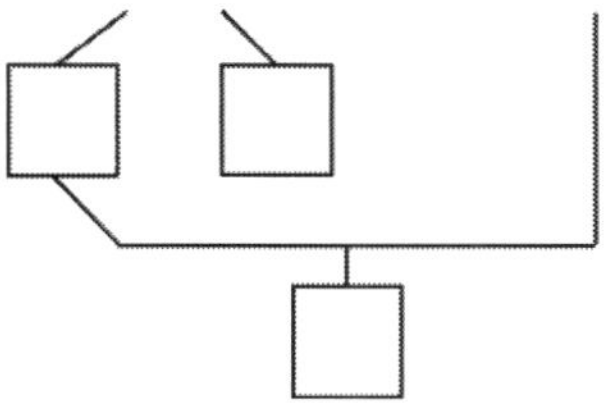

42. Calculate the following questions.

| | |
|---|---|
| 8 – 3 = | 9 – 6 = |
| 48 – 3 = | 59 – 6 = |

| | |
|---|---|
| 40 – 30 = | 70 – 20 = |
| 45 – 30 = | 75 – 20 = |

**43.** In the seafood market. There were 78 fish in the water tank. Customer bought some fish and 28 fish left in the water tank. How many fish does the customer buy from the water tank?

**44.** Victor and Mia are collecting empty bottles to protect the environment. Victor collected 76 bottles, which is **5 more than Mia**. How many bottles did Mia collect?

**45.** Kiki and Wendy received some red packets in the Chinese New Year. Kiki received 46 dollars, and it is 6 dollars more than Wendy's. How much money did they received altogether?

## Module 13 Practice Lesson 4

35. Find the missing numbers in the following questions. The first question has been done for you.

73 − 5 = [68]

[60] [13]

[8]

42 − 4 = [ ]

[ ] [ ]

[ ]

74 − 8 = [ ]

[ ] [ ]

[ ]

53 − 9 = [ ]

[ ] [ ]

[ ]

36. Circle the sticks that you need to take away and calculate the answer.

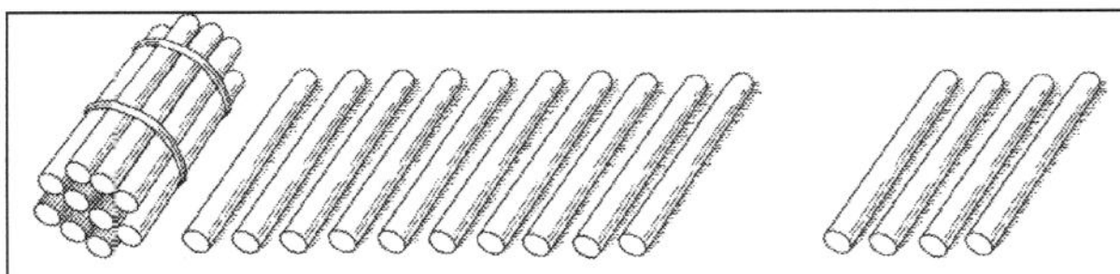

[24] − [7] = [ ]

First calculate:

[ ] − [ ] = [ ]

Then calculate:

[ ] + [ ] = [ ]

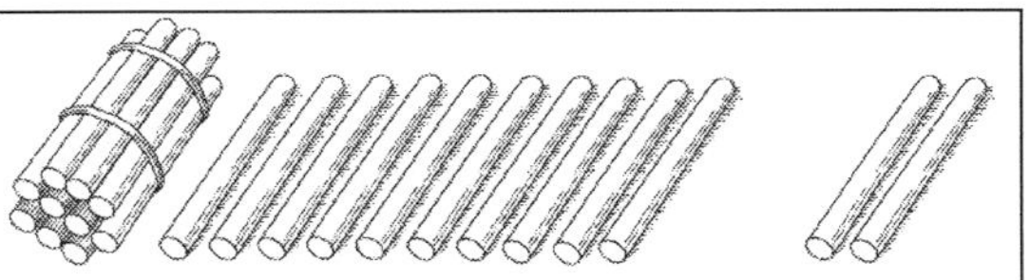

[22] − [6] = [ ]

First calculate:

[ ] − [ ] = [ ]

Then calculate:

[ ] + [ ] = [ ]

37. Compare each number sentence by using “>”, “<” or “=”.

| | | |
|---|---|---|
| 45 + 8 ◯ 64 − 7 | 51 − 4 ◯ 38 + 9 | 34 − 7 ◯ 17 + 8 |
| 24 + 20 ◯ 39 + 9 | 42 + 8 ◯ 57 − 7 | 56 − 7 ◯ 38 + 9 |
| 38 + 7 ◯ 54 − 7 | 34 + 7 ◯ 45 − 8 | 27 + 9 ◯ 45 − 6 |

38. Benjamin bought a story book with 74 pages in total. He read 4 pages on the first day and 5 pages on the second day. How many pages are left unfinished?

39. Three students are planting trees in order to protect the environment. Aaron planted a total of 74 trees, and Benjamin planted 6 fewer trees than Aaron. Calvin had planted 9 more trees than Benjamin. How many trees did each of them plant?

Aaron planted __________ trees,
Benjamin planted __________ trees,
Calvin planted __________ trees.

40. Henry has 72 dollars in total. He bought a 9 dollars burger and a 20 dollars family meal. How much money does he have left after the meal?

## Module 14 认识钱币 Introduction to money

There are two types of money in Australia. A coin is a piece of flat rounded metal with some special stamps on it. A note is a piece of plastic paper with different colours representing different amounts of money.
We use the symbol *c* to represent cents, and $ to represent dollars.

### ❖ 硬币的认识 Introduction to coins

There are 6 coins of different values in Australia, these are 50 cents, 20 cents, 10 cents, 5 cents, 2 dollars and 1 dollar.

In Australia, the cents coins are silver, and the dollar coins are gold.

### ❖ 钞票的认识 Introduction to notes

There are 5 notes of different values in Australia, these are 5 dollars, 10 dollars, 20 dollars, 50 dollars and 100 dollars.

| | | | | |
|---|---|---|---|---|
| $ 5 in pink | $ 10 in blue | $ 20 in red | $ 50 in yellow | $ 100 in green |

**Think:** *Do you know why we don't have $15 or $23 notes?*

## Module 14 Practice Lesson 1

41. What is the sign for cent?

42. What is the sign for dollar?

43. Find how much money are in the following diagrams.

There are ______________ cents in total

There are ______________ dollars in total

There are ______________ dollars in total

There are ______________ cents in total

44. Find how much money are in the following diagrams.

There are ______________ dollars in total

There are ______________ dollars in total

There are ______________ dollars in total

There are ______________ dollars in total

45. How many of the following items can you buy if you have 50 cents?

30 cents | 40 cents | 50 cents | 10 cents

46. According to the size of each cent coin, find in the correct value for them.

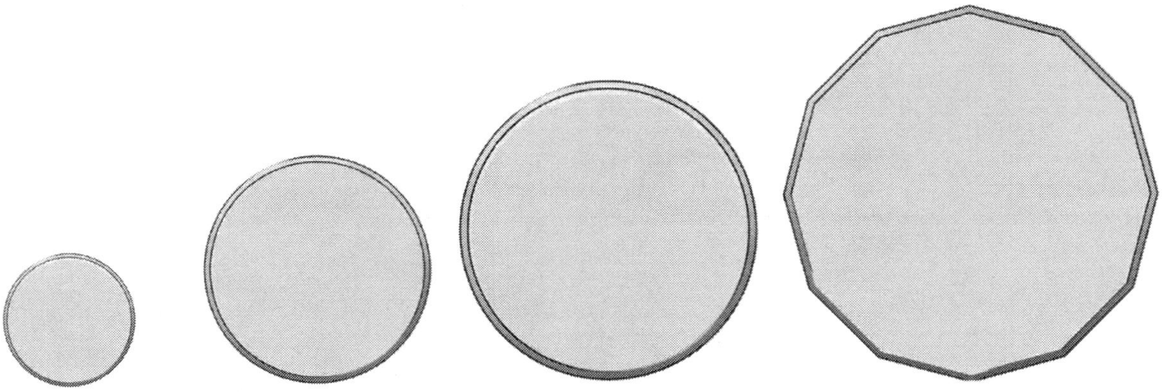

47. Order the notes from the largest to the smallest value using numbers 1 to 5.

________ ________ ________ ________ ________

48. Order the notes from the smallest to the largest value using numbers 1 to 6.

________ ________ ________ ________ ________ ________

## Module 14 Practice Lesson 2

28. Find how much money are in the following diagrams.

| Australian coins and notes | Total amount ($ or c) |
|---|---|
|  | |
|  | |
|  | |
|  | |
|  | |
|  | |

29. State whether the following statements are true or false.

a) There are two types of money in Australia, coins and notes. (True/False)

b) The larger the size of the coin mean the larger the value. (True/False)

c) Dollar coins are gold and cents coins are silver. (True/False)

d) We can use Australian money in another countries. (True/False)

e) We have 1 cent coin in Australia. (True/False)

f) $5 notes are in pink colour and $10 notes are in blue colour. (True/False)

g) If I have a $1 coin and a 50c coin, this is $51 in total. (True/False)

h) In Australia, coins are made from plastic and notes are made from paper. (True/False)

i) In Australia, the smallest money is 5c and largest money is $100 (True/False)

30. Complete the following addition number sentences for each diagram below. The first question has been done for you.

50 + 20 + 5 = 75 c

□ + □ + □ = □

□ + □ + □ = □

□ + □ + □ = □

31. Complete the following subtraction number sentences for each diagram below. The first question has been done for you.

57 – 5 – 2 = $50

□ – □ – □ = □

□ – □ – □ = □

□ – □ – □ = □

## Module 14 Practice Lesson 3

46. Connect.

Rem buys a

for 35 c.

Kira buys a

for 45 c.

Vivi buys a

for 80 c.

Hyo buys a

for 75 c.

Yumi buys an

for 90 c.

Gina buys a

for 60 c.

Dan buys a

for 55 c.

Peter buys a

for 85 c.

47. Jessica wants to buy a notebook for $25 and a pencil bag for $30. How much money does she need to pay altogether?

48. Aaron has $20. he wants to buy a pair of sneakers for $85. How much more money does he need to save to buy the sneakers?

49. Gina wants to buy 5 toy cars for her son and daughter. If each toy car is $11, how much money does she need to pay in total?

50. Today is Nina's birthday. Her father gives $10 to Nina and her mother gives her $8. After that, her grandparents also gives another $9 to her. How much money does Nina have in total?

51. Fiona wants to buy some flowers for her mother. If Fiona buys three roses and one lily. How much money does she need to spend?

| Rose | Lily |
|---|---|
| $5/each | $7/each |

## Module 14 Practice Lesson 4

41. How much change will each student get? Connect the question to the correct answer.

Tina has $20 and buys a  for $8.

Aidan has $30 and buys a  for $7.

Joelle has $18 and buys a  for $8.

Bella has $14 and buys a  for $5.

Joshua has $50 and buys a  for $20

Edward has $20 and buys a  for $5.

Isaac has $15 and buys a  for $11.

Lisa has $10 and buys a  for $5.

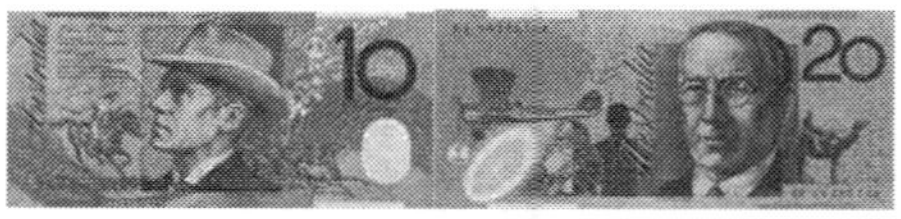

42. Cecilia buys a story book for $20. She pays $50 to the cashier. How much change will he get back?

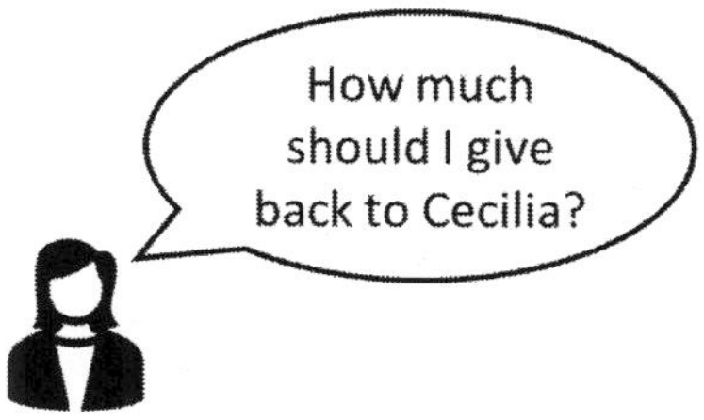


43. Oscar buys a pack of playing cards. He gives $20 to the cashier and receives $9 as change. How much does a pack of playing cards cost?

44. Henry can save $5 every day. How much money can he save in a week?

45. Johnny has a bag of coins. He draws three coins from the bag.
The first coin is silver colour, and it has the largest size.
The second coin is also silver colour, and it has the smallest size.
The last coin is a gold colour, and it has the largest value in all these coins.

What are these three coins? And how much are they in total?

## Module 15 概率的认识 Introduction to probability

Probability is how likely something is to happen. In our daily lives, there are lots of events that cannot be predicted. For example, you cannot say with certainty that tomorrow will rain. Then how can we describe something that might happen in the future?

### ❖ 可能性的认识 Introduction to chance

Chance has the same meaning as probability. It can describe how likely something is to happen. There are some useful chance words you might need to remember:

Certain, always happen, must happen, very likely, sometimes happen, likely, even chance, very unlikely, never happen, impossible, possible, maybe, might etc.

Now, try to describe some events with your classmates that will always happen, sometimes happen and never happen.

| | |
|---|---|
| Always happen | |
| Sometimes happen | |
| Never happen | |

### ❖ 概率的表示 Probability on a number line

Probability can be represented in numbers between 0 and 1. In earlier schooling we don't need to worry too much about representing probability by using numbers. However, you need to understand the probability scale from most likely to least likely.

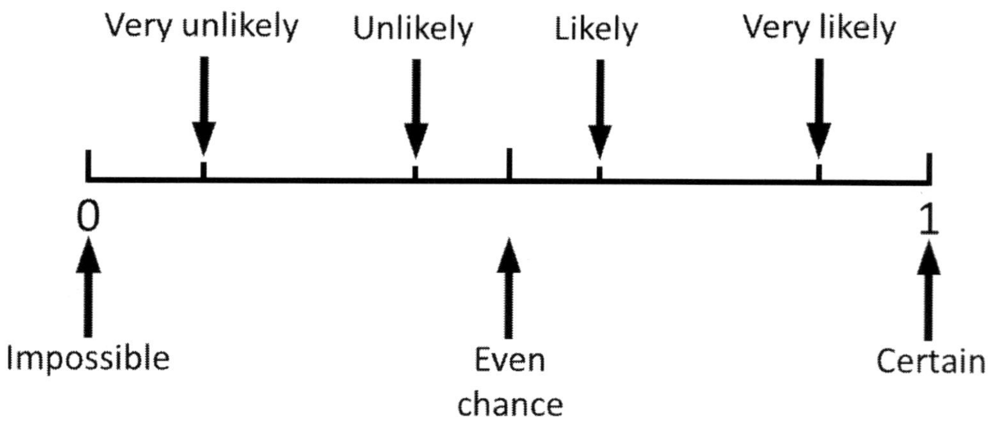

## Module 15 Practice Lesson 1

49. Tick the box which best describes the following events.

| | Impossible | Certain | Maybe |
|---|---|---|---|
| I will have my breakfast tomorrow morning. | | | |
| It will rain tomorrow. | | | |
| Tomorrow is my birthday. | | | |
| I need to go to school tomorrow. | | | |
| I can sleep very late tonight. | | | |
| I need to do my homework tonight. | | | |
| My mother will pick me up after school. | | | |
| I will walk home today. | | | |
| My father will give me $200 tomorrow. | | | |

50. There are some marbles in the box. Colour these marbles to match with the statement below.

| | |
|---|---|
| It is certain to pick a red marble 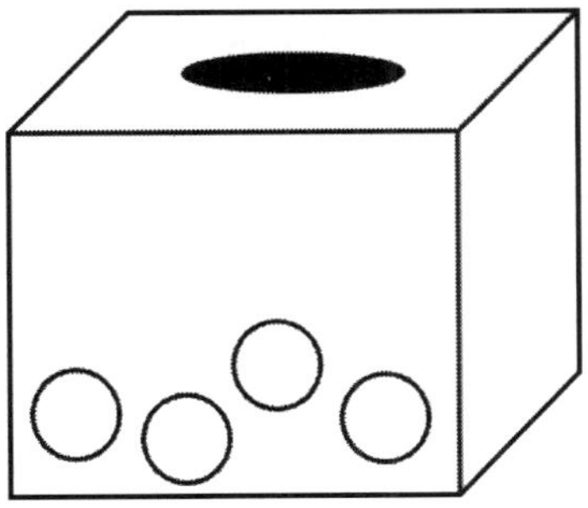 | It is impossible to pick a red marble 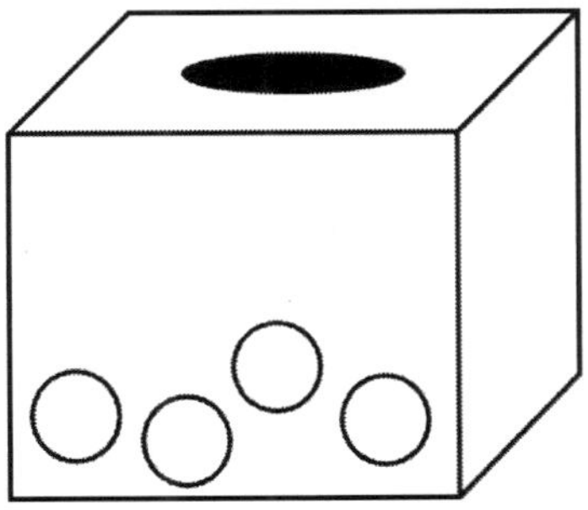 |
| It is unlikely to pick a red marble 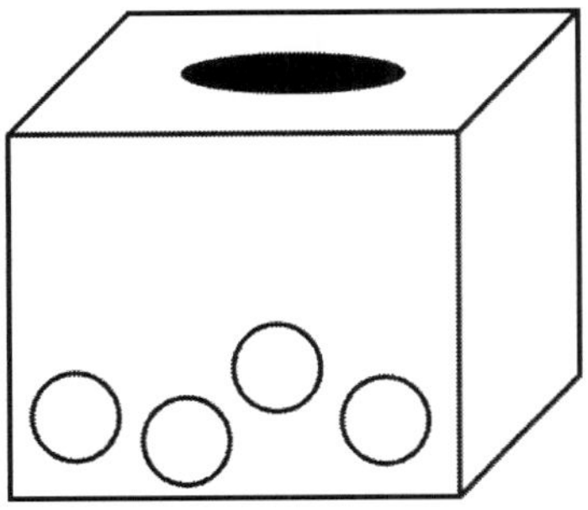 | It is most likely to pick a red marble 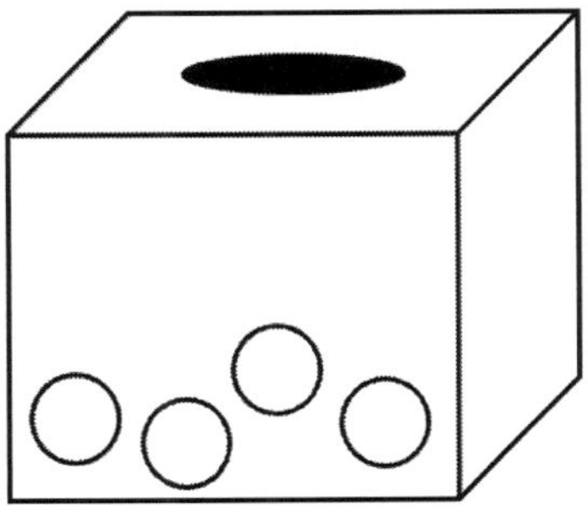 |

51. Tick the following chances that the spinner are going to land on shaded area.

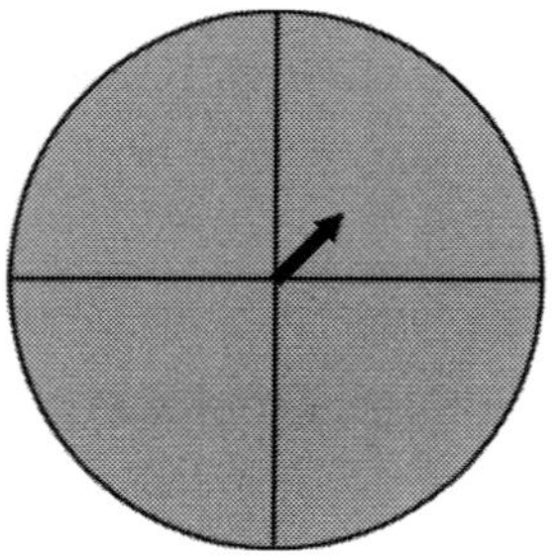

Certain / Even chance / Most likely / Impossible

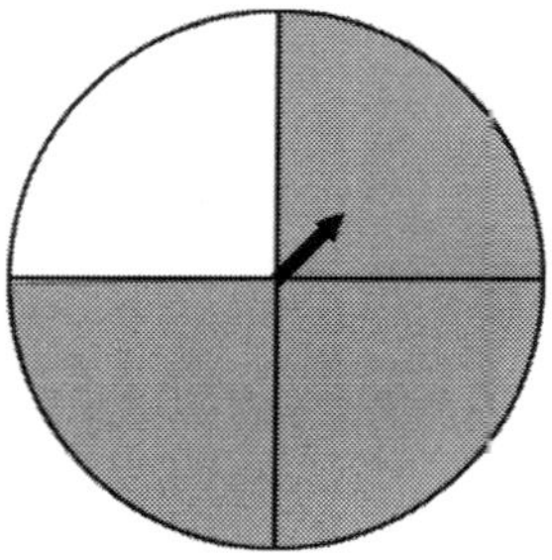

Certain / Even chance / Most likely / Impossible

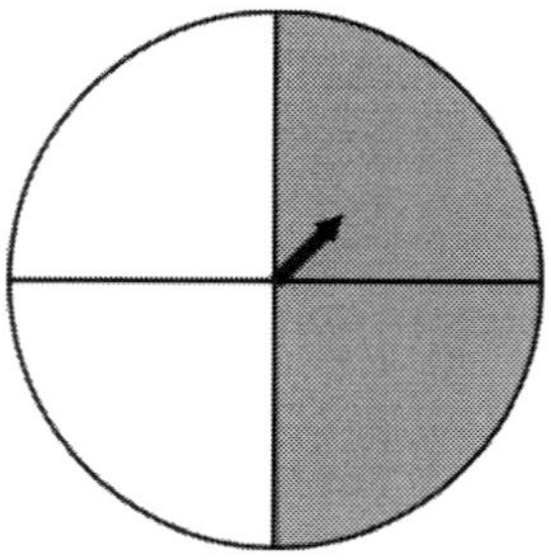

Certain / Even chance / Most likely / Impossible

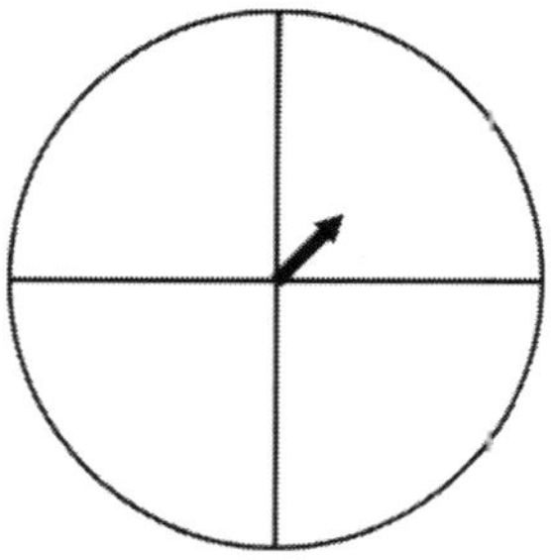

Certain / Even chance / Most likely / Impossible

52. Order the following events from the least likely to most likely by using number 1 to 4.

53. Write the chance word to identify the following events. (Impossible, certain, likely, unlikely)

 I will go to school on school day

 I will have music class tomorrow

 An alien student will come to my class tomorrow

 I will have math class tomorrow

## Module 15 Practice Lesson 2

32. We have already learnt about money in the previous chapter. We know that coins have two faces, head and tail. Please do the following experiment with your teacher and classmates.

a) When toss a coin in the air, do you think which face will come up when drop on your hand?

Head Tail

b) Toss the coin 10 times to see which face has greater probability of coming up. Use 'H' for head and 'T' for tail.

| 1 | 2 | 3 | 4 | 5 | 6 | 7 | 8 | 9 | 10 |
|---|---|---|---|---|---|---|---|---|---|
| | | | | | | | | | |

__________ heads come up.

__________ tails come up.

c) From the results in the previous part, can we say that the one probability of getting head or tail (based on your answer) must be greater? Why?

d) Now, this time toss the coin 30 times to see which face has greater probability of coming up. Use tally to record your data.

| Faces | Tally | Total |
|---|---|---|
| Head | | |
| Tail | | |

e) What do you think would happen if I toss the coin 100 times? Which side do you think is more likely to be facing up?

## Module 15 Practice Lesson 3

1. If you pick up a marble at random without looking, which colour are most likely to be picked?

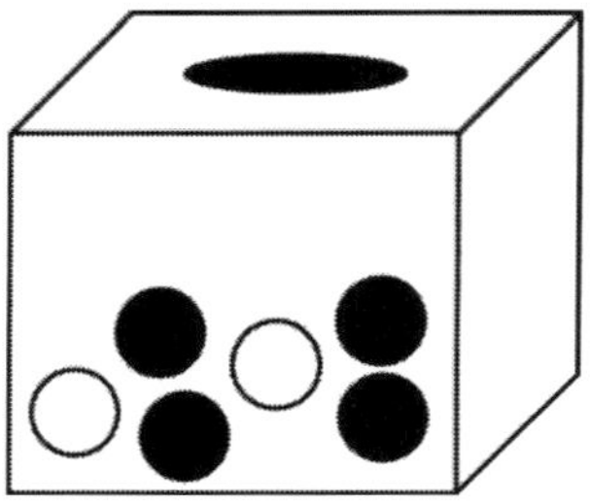

Black

White

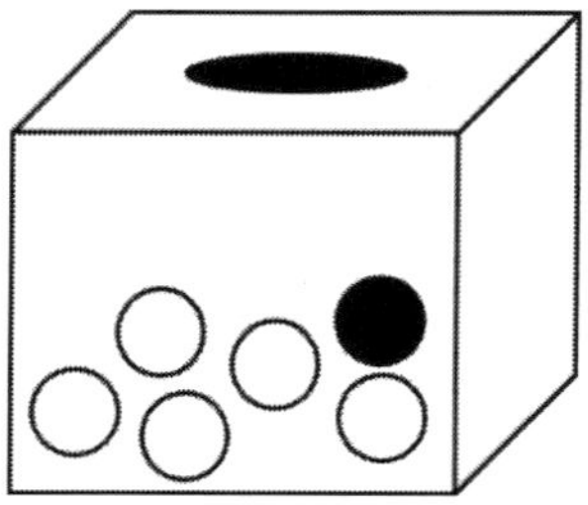

Black

White

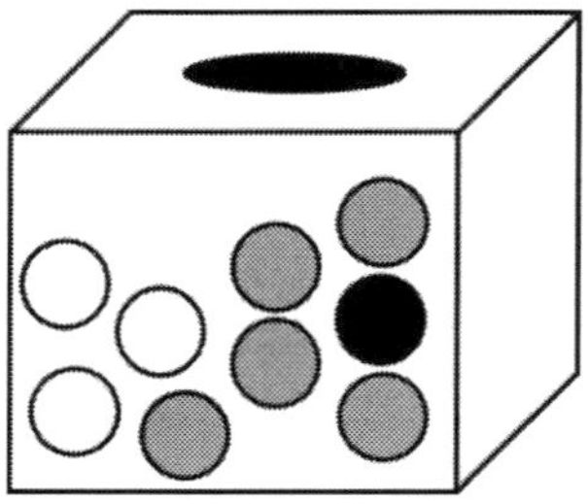

Black

White

Grey

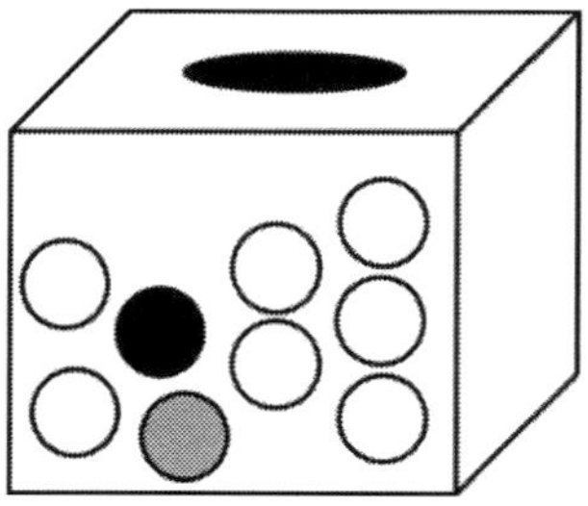

Black

White

Grey

2. Circle the colour that is most likely to be selected by the spinner below?

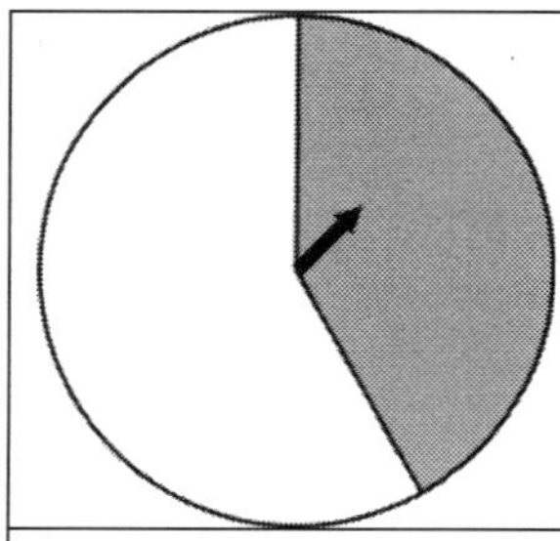

White

Grey

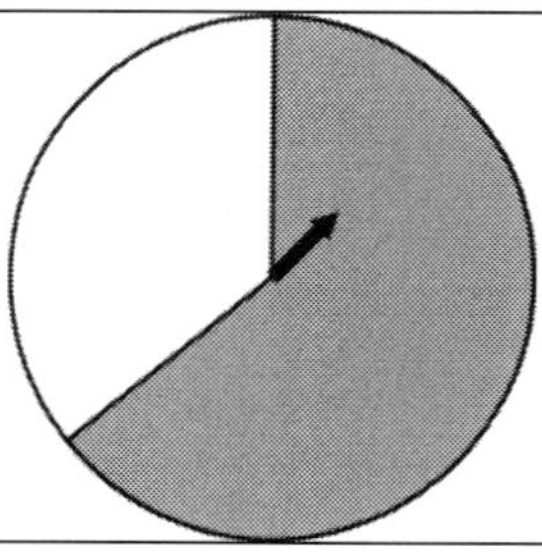

White

Grey

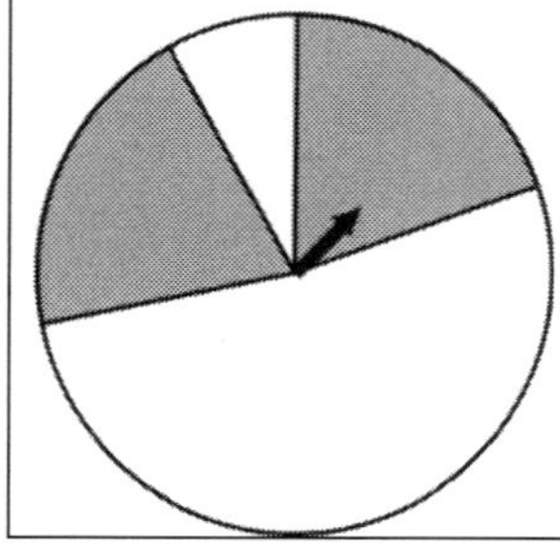

White

Grey

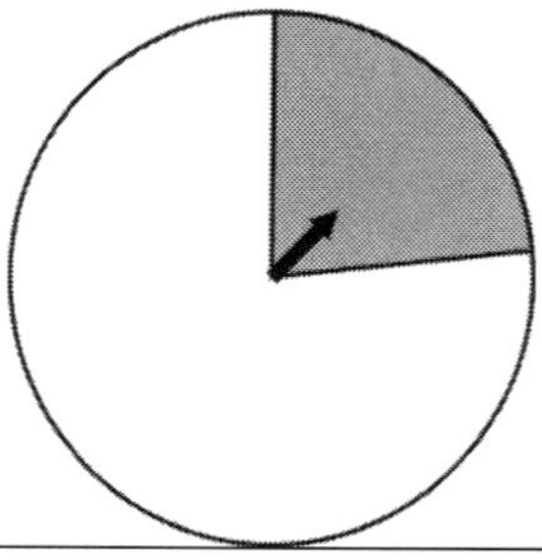

White

Grey

3. Colour the following spinners to make each sentence true.

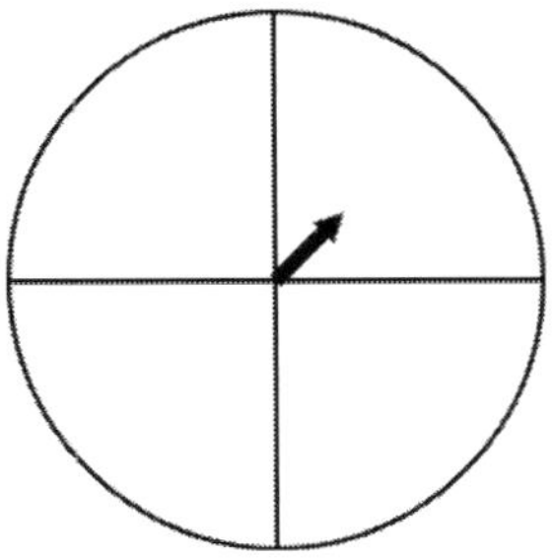

It is certain to have a black colour.

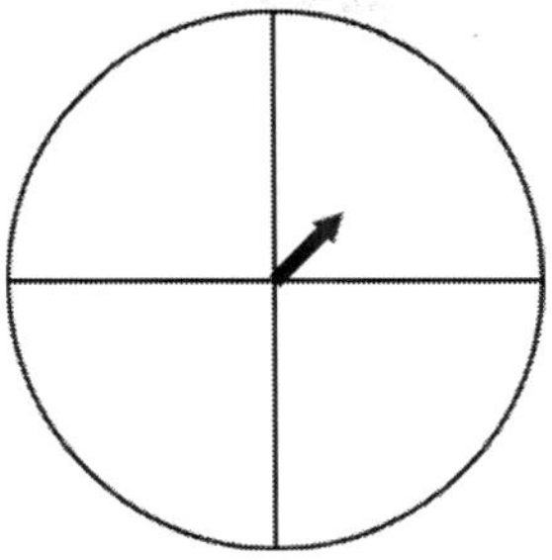

It is impossible to have a black colour.

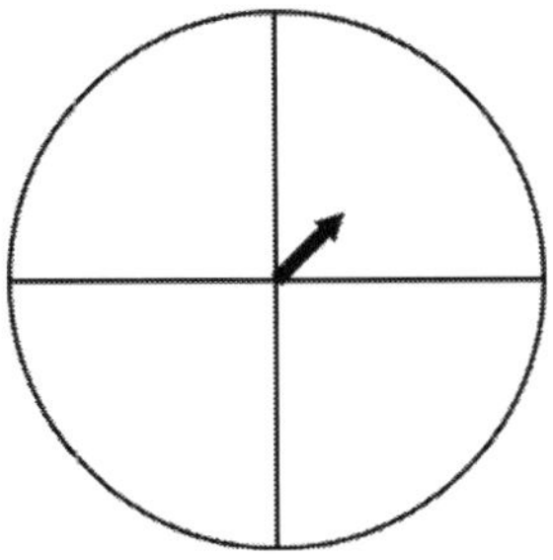

It is unlikely to have a black colour.

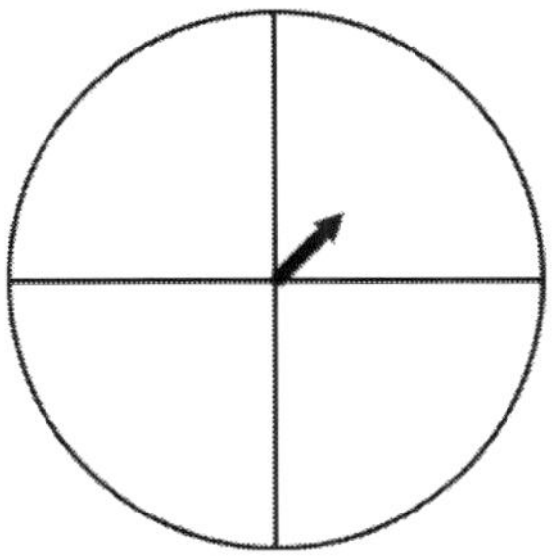

It is most likely to have a black colour.

4. There are some shapes in the box. Draw those shapes to make each sentence true.

Please use the following shapes to complete these questions.

It is most likely to pick a heart shape.

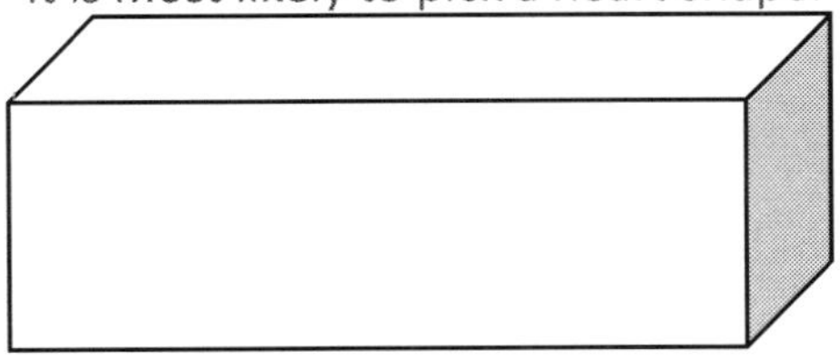

It is less likely to pick a triangle shape.

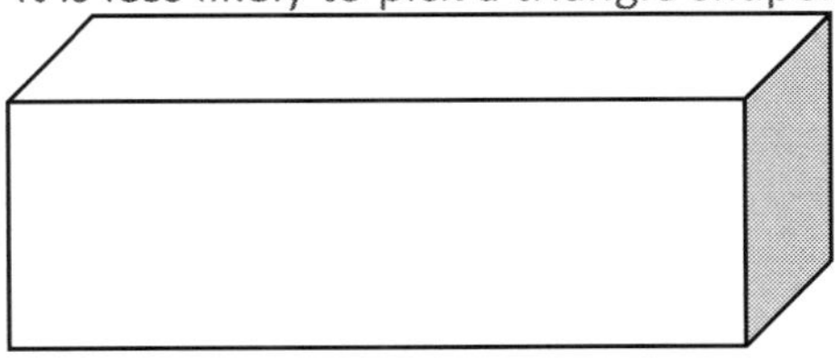

It is certain to pick a circle shape.

It is impossible to pick a star shape.

The probability of getting a star is greater than getting a circle.

The probability of getting a triangle is lower than getting a heart.

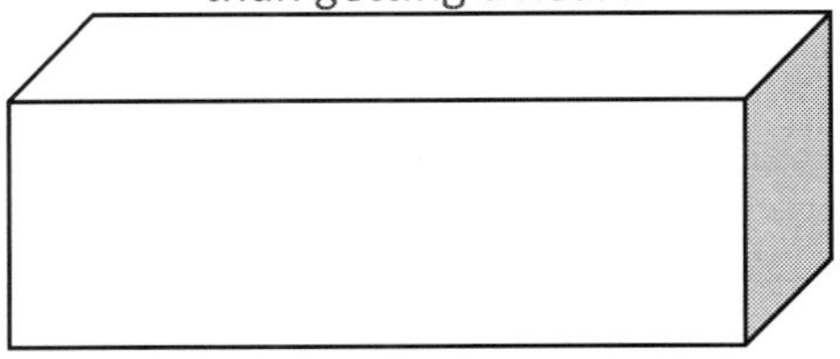

## Module 15 Practice Lesson 4

46. Do you guys know the game Monopoly? In that game, players need to roll a die to decide how many moves they can make.
A normal die has six sides from 1 to 6. Please do the following experiment with your teacher and classmates.

a) When roll one die, which number do you think will come up with the greatest chance?

b) Roll the die 10 times to see which number has greater probability of coming up.

| First | Second | Third | Fourth | Fifth | Sixth | Seventh | Eighth | Ninth | Tenth |
|---|---|---|---|---|---|---|---|---|---|
| | | | | | | | | | |

"1" comes up ________ times
"2" comes up ________ times
"3" comes up ________ times
"4" comes up ________ times
"5" comes up ________ times
"6" comes up ________ times

c) From the result of the previous question, can we say that the probability of getting one number must be greater than getting another number? Why?

d) Now, this time roll the die 50 times to see which number has greater probability of coming up. Use tally to record your data.

| | Tally | Total |
|---|---|---|
| 1 | | |
| 2 | | |
| 3 | | |
| 4 | | |
| 5 | | |
| 6 | | |

e) What do you think would happen if I toss the coin 100 times? Which side do you think is more likely to be facing up? Discuss with your classmates.

## Module 16 平均分 Equal shares

Tonight, your family will be having a pizza as dinner. If you and your sibling need to share four slices of pizza. How will you share these four slices?
On dividing the whole or a group of items is into equal parts, we get **equal shares**.

### ❖ 探索半分 Exploring halves

Halves is splitting a whole or a group of items into two equal parts.

Half of a lemon

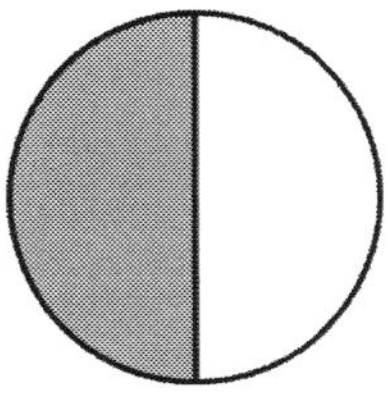

Half of a circle

Half of a jug

Half of 4 slices of pizza

### ❖ 平均分技巧 Sharing strategies

In our real word, we can share a whole or a group of items more than two equal parts. Such as share into group of three or group of four. In our early schooling, we only need to know how to share into two parts and four parts.

| | |
|---|---|
| 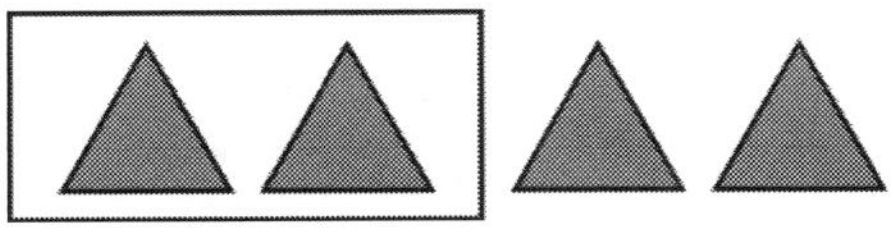<br>Half of 4 triangles are 2 triangles | 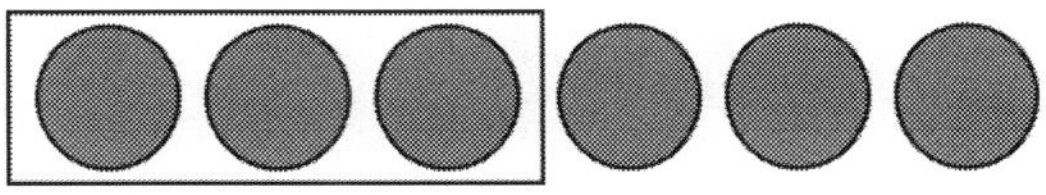<br>Half of 6 circles are 3 circles |
| <br>Half of 8 hearts are 4 hearts | <br>Four parts of 8 stars are 2 stars |

## Module 16 Practice Lesson 1

**54.** Tick the equal shares of the following diagrams.

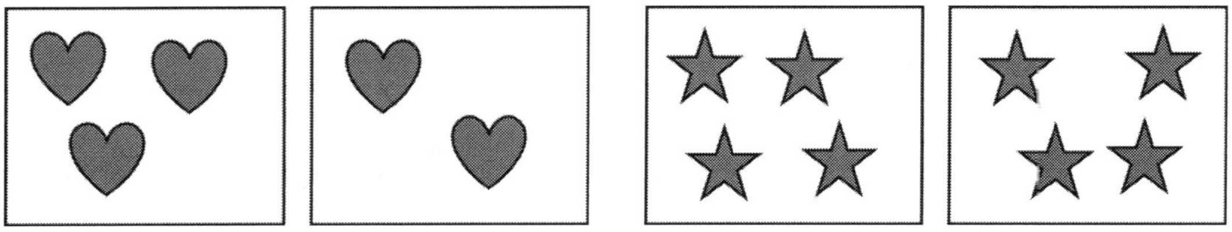

( )　　　　　　( )

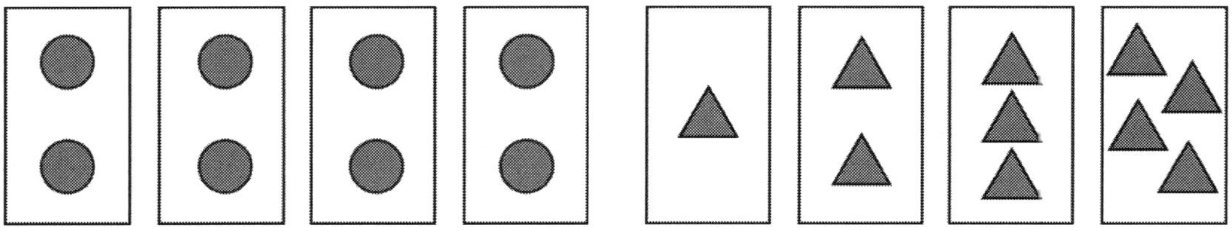

( )　　　　　　( )

**55.** Draw the missing shapes to make them become equal shares.

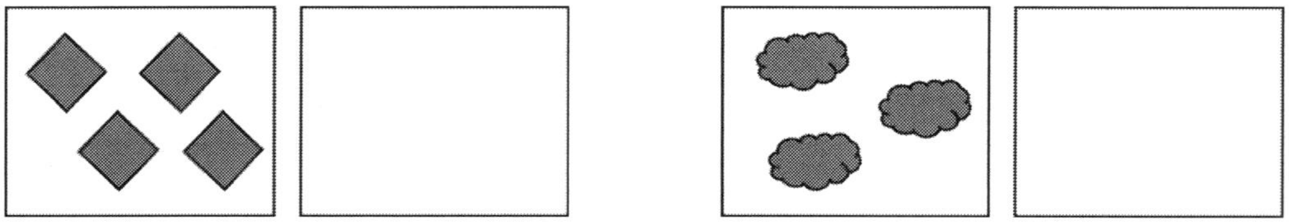

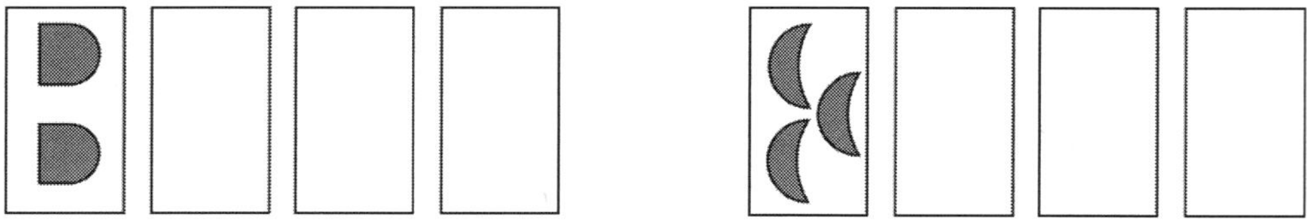

**56.** Mr. Chen has 10 apples. He wants to share those apples between 2 of his friends. How many apples does each friend get?

## Module 16 Practice Lesson 2

33. Circle the correct number of items for the following questions.

| | |
|---|---|
| Divide the clock into 2 groups  | Divide the hammers into 4 groups  |
| Divide the cats into 2 groups  | Divide the dogs into 4 groups  |

34. Connect the groups into correct statement.

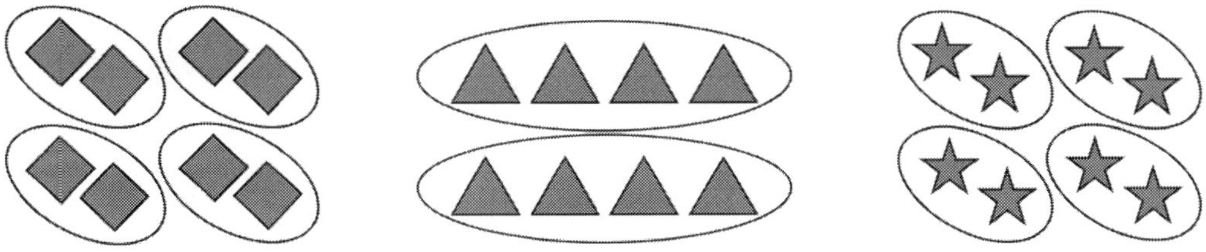

Groups of 2

Groups of 4

35. Halve the following shapes and shaded it.

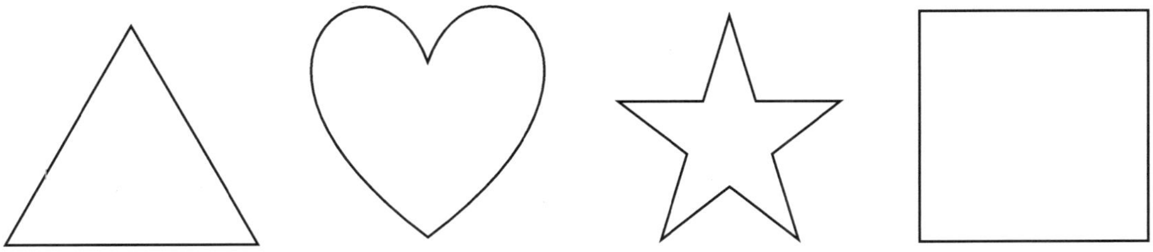

36. Johnny has 8 cups. He wants to put those cups into 2 boxes. If he want to make sure each box has the same number of cups, how many cups he needs to put in each box?

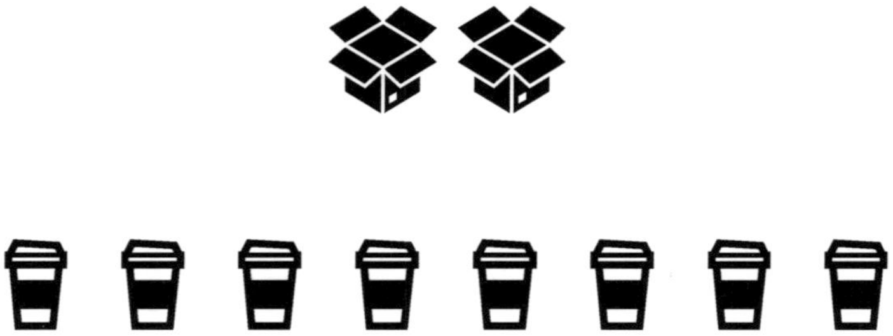

37. Fiona has 12 apples. She wants to put those apples into 4 plates. If she wants to make sure each plate has the same number of apples, how many apples are in each plate?

38. Joe has 2 boxes of marbles. If each box has 9 marbles, how many marbles are there altogether? Draw those marbles in each boxe.

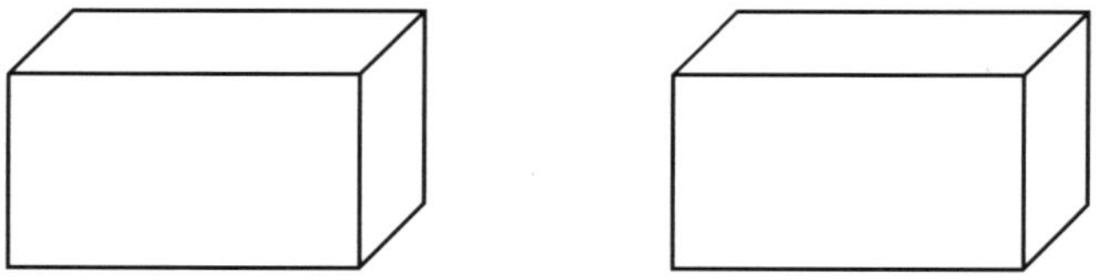

39. Dan has 4 pencil holders. If each pencil holder carries 5 pencils, how many pencils are there altogether? Draw those pencils in each pencil holder.

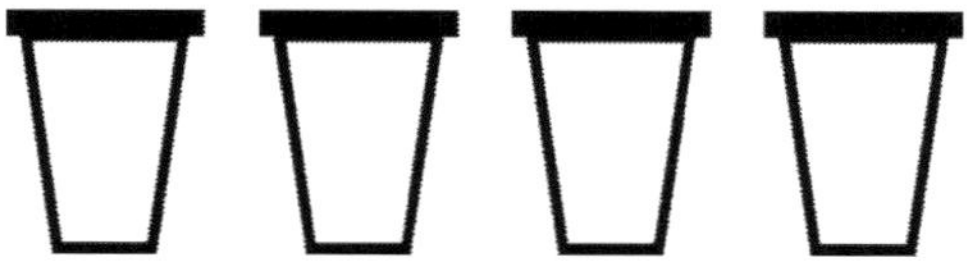

## Module 16 Practice Lesson 3

**5.** There are 24 marbles. Please split these marbles into equal parts by circle them.

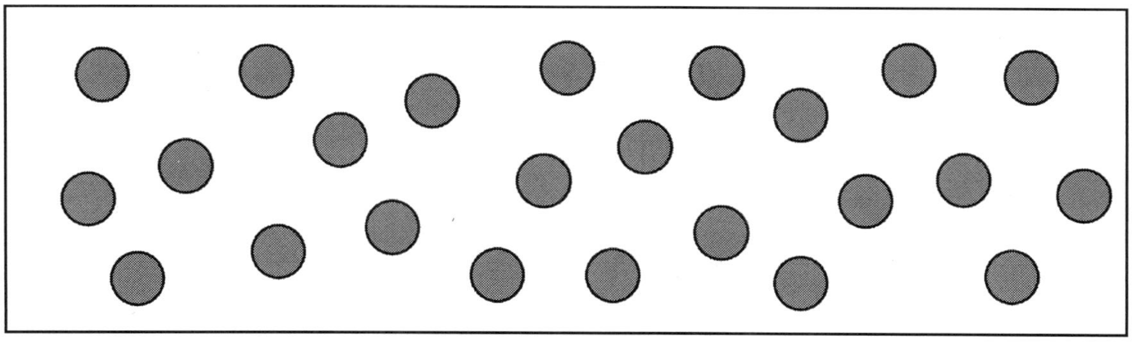

a) How many groups of marbles you circled?

b) How many marbles are in each group?

6. Circle the following shapes that have been split into equal parts.

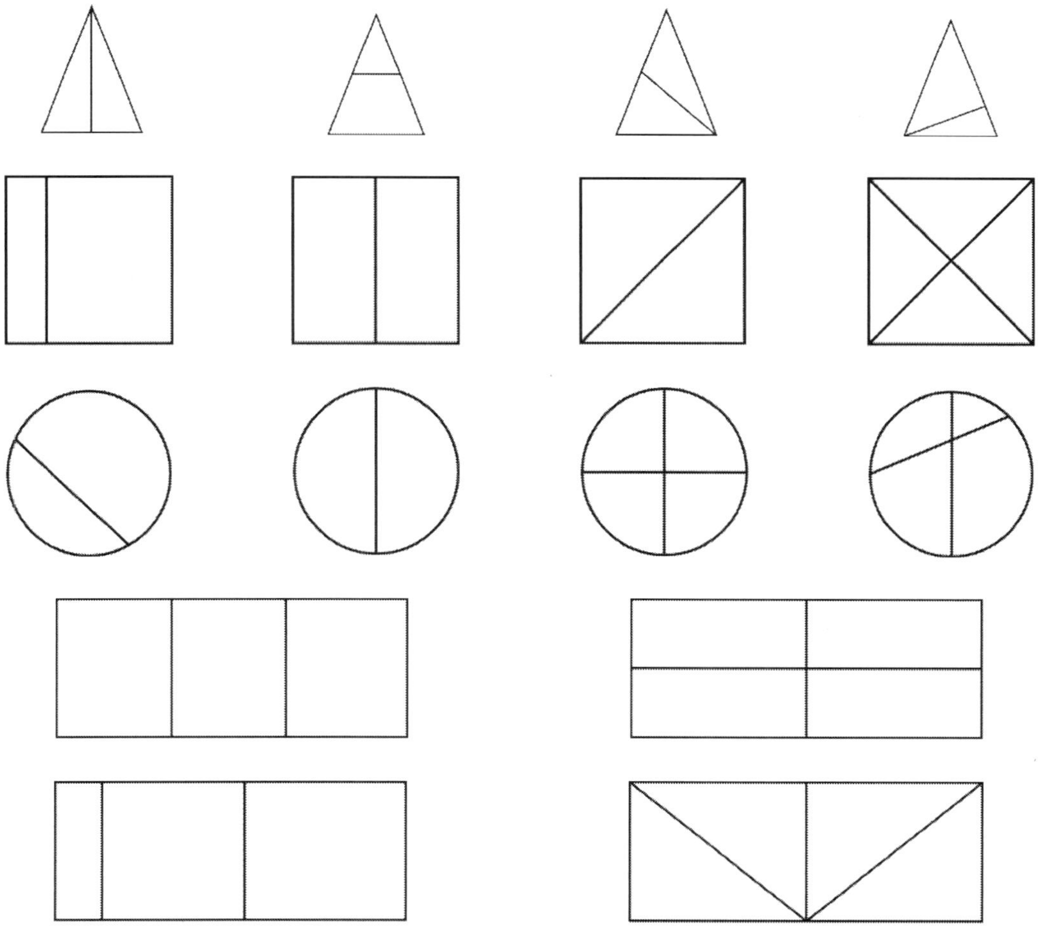

## Module 16 Practice Lesson 4

47. Mr. Yu has put out the apples for three students to eat. Look at the diagram, is there anything wrong?

Can you rearrange the apples in each box so that each student gets an equal amount?

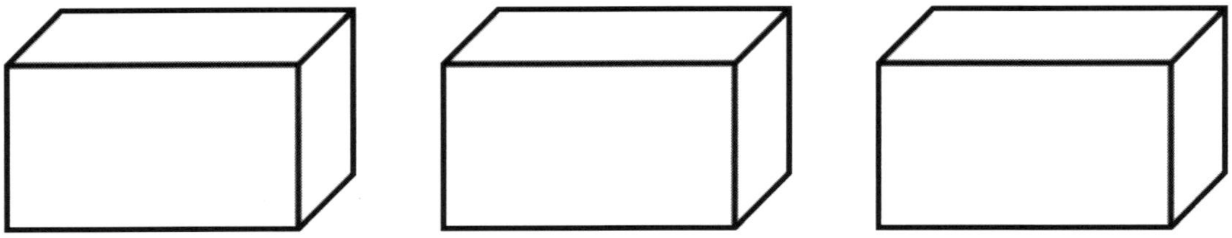

48. Henry ordered 20 burgers. He wants to split them into group of 2. How many groups can Henry have?

49. Split the following shapes into four equal shares

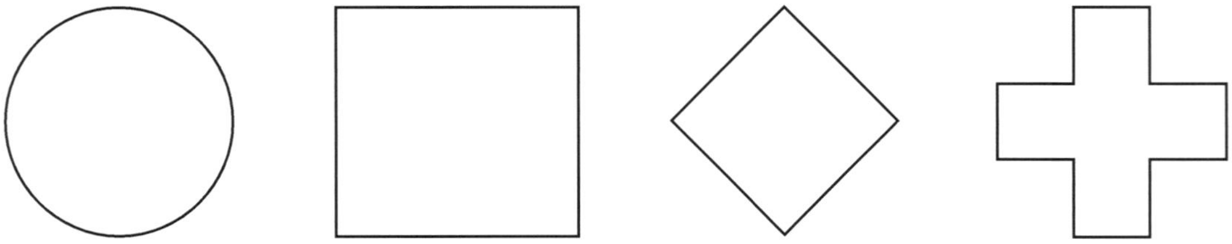

50. There are some black marbles and white marbles. Please shade some white marbles to make both colours have the same amount.

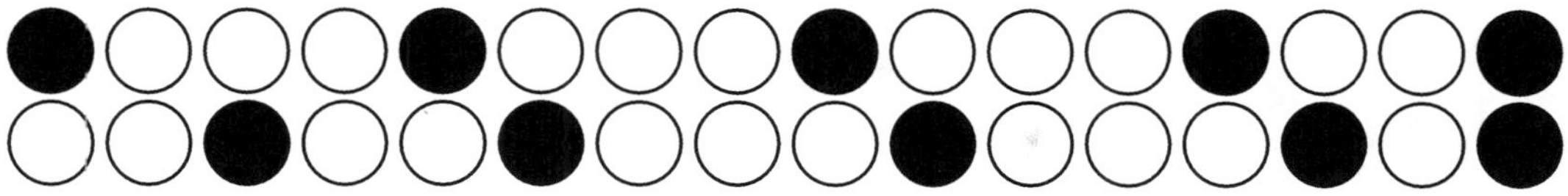

51. Jess has 14 books. If each bag can only carry 2 books. How many bags does she need?

52. Josh wants to put 9 apples into some bowls. If each bowl can hold 2 apples, how many apples are left out? How many bowls does he need?

53. Year 1 students are having an excursion. There are 15 students. If each car can carry 4 students. How many cars do they need?

54. How many students are into your classroom? If students are having a group of 2, how many groups are there in total? And how many students are left out?

Made in United States
Orlando, FL
04 November 2026

90207801R00081